LOVE THEM TO DEATH

LOVE THEM TO DEATH

Turning Invasive Plants into Local Economic Opportunities

Edited by **WENDY L. APPLEQUIST**

MISSOURI BOTANICAL GARDEN PRESS

Disclaimer: This book discusses a variety of potential uses of invasive species. Anyone wishing to implement similar uses in practice will need to do additional research. It is essential that any plants collected for human or animal consumption be correctly identified; this book does not include aids to botanical identification. General information regarding traditional medicinal uses of plants does not include necessary details such as preparation, dose, or contraindications, and does not constitute medical advice. If you have questions about medicinal use of plants for an illness, please consult a health care professional who is knowledgeable about that subject. This book cannot provide legal advice regarding federal, state, or local regulations that may pertain to the harvest, possession, transport, manufacture, or sale of invasive species or products made from them. The reader is responsible for ensuring that any activities engaged in are safe and legal; the authors, editor, and Missouri Botanical Garden Press bear no responsibility for uses of the information presented.

ISBN 978-1-935641-31-5
Library of Congress Control Number 2024936683

Managing Editor: Allison M. Brock
Editor: Lisa J. Pepper
Press Coordinator: Amanda Koehler
Cover and interior design: Kim Scott/Bumpy Design

Copyright © 2025 by Missouri Botanical Garden Press
4344 Shaw Boulevard, St. Louis, Missouri 63110, U.S.A.
www.mbgpress.org
All rights reserved. Printed in U.S.A.

Front cover photo: teasel by Theresa Hornstein. Back cover photos (left to right): *bon* soup by V. D. Nguyen; buckthorn berries on wool by Theresa Hornstein; student with bittersweet basket by Katie Grove; trained cows in weedy pasture by Kathy Voth.

Contents ◆ ◆ ◆

Acknowledgments..vii
Introduction...1

CHAPTER 1   When Kudzu Ate the South, Why Didn't the South Eat Kudzu?.....9
            Katie Carter King

CHAPTER 2   Welcome to the Green Food Zone: Wild Taro
            (*Colocasia esculenta*) 19
            Peter J. Matthews, Mohammad Anwar Hossain & Van Dzu Nguyen

CHAPTER 3   A Brief Biography of Itadori: Re-storyation as an
            Approach to Japanese Knotweed Management 50
            Tusha Yakovleva

CHAPTER 4   Cooking with Invasive Species: Culinary Suggestions
            for Promoting Control through Consumption 66
            Alana N. Seaman & Alexia Franzidis

CHAPTER 5   Can We Love Invasive Species to Death?........................80
            Sara E. Kuebbing, Joshua Ulan Galperin & Martin A. Nuñez

CHAPTER 6   Turn Your Livestock into Weed Managers.......................112
            Kathy Voth

CHAPTER 7   Invasive Plants Used in Chinese Medicine.....................130
            Thomas Avery Garran

CHAPTER 8   Some Invasive Species of Demonstrated Medicinal Value 163
            Wendy L. Applequist

CHAPTER 9   Invasive Common Reed as Valuable Bio-Resource:
            Lessons Learned from Europe...............................188
            Franziska Eller

CHAPTER 10 Remediation as Harvest: Invasive Plant Species
            as Building Materials 216
            Katie MacDonald & Kyle Schumann

CHAPTER 11 An Ethic of Care in Basketry: Weaving with Invasive Vines.......228
            Katie Grove

CHAPTER 12 Invasive Color: Using Invasive Species as Natural Dyes 241
            Theresa Hornstein

CHAPTER 13 Fiber Optics: Do Invasive Species Look Good on Paper?.........253
            James Ojascastro

Contributors...291

Acknowledgments ◆ ◆ ◆

The Missouri Botanical Garden Press and its staff, in particular Lisa Pepper, are gratefully thanked for their support for this project and careful editing of manuscripts, and Kim Scott for the design and layout of the book.

Peer reviewers who read and commented on draft chapters, and supplied many useful suggestions, are gratefully acknowledged. These include Matthew A. Barnes, Timothy Barrett, Brenda Frick, Stefan Gafner, Robbie Hart, Justin Holt, Sarah Howard, Billie Lee, Shawn McCourt, Erin Moore, Thomas Mozdzer, Anke Nordt, Fred Provenza, Bill Schoenbart, Megan Singleton, Mel Sweetnam, Matt Tommey, Roy Upton, and Marc Williams, as well as those who did not wish to be named but are equally appreciated.

This project was sparked by a conversation with the editor's late husband, Bob Fuqua, and would not have existed without his inspiration and encouragement.

Introduction

Wendy L. Applequist

Invasive species are an increasing ecological and economic problem in the modern era, when global transport has spread many exotic species that can harm local ecosystems. These include pests and pathogens, from insects to bacteria and fungi that can infect animals or plants, as well as more conspicuous plant and animal species. A recent, detailed and deliberately conservative estimate of the dollar value of harms done by invasive species in the U.S. (Fantle-Lepczyk et al., 2022) placed it at $21 billion per year in the decade from 2010 to 2020, with a total cost between 1960 and 2020 of at least $1.22 trillion. Agriculture was the most affected sector, at over half a trillion dollars. Except for the broad "other" category, which were mostly disease-causing microorganisms, the costliest harms were caused by introduced mammals, insects, and plants.

The scientific and cultural status of so-called "invasive plants" in particular is a controversial topic, which will not be delved into here. Many introduced plants coexist with native plants in new habitats without doing obvious harm. Some authors have argued that calling them "invaders" just because they are not native conveys needless fear and hostility, and may inspire control methods that do more harm than good. The term "invasive" is even sometimes used for native plants that farmers find inconveniently weedy. Still, some plants really can do serious damage in new habitats, where the pests, diseases, competitors, or environmental challenges that once limited their growth are absent. Woods choked by masses of kudzu or Japanese honeysuckle, or lakes covered with water hyacinth, provide visual evidence for lovers of nature that something is wrong.

At the same time, there's an increasing feeling that the economy is not working well for many people, who might benefit from access to more diverse sources of income or useful resources. In the past, the living world provided us with almost all of the resources we used, from food to fuel, timber, fibers, medicines, and so forth. Once, most of these goods were obtained from plants we would call "wild," though they were often carefully managed to ensure their availability. Today, most plant products we use are cultivated, even trees for paper and timber production. The labor required for wild harvesting usually makes it uneconomical, except for high-value products or those that can't readily be cultivated. Moreover, intact land and public access to land are both limited, and when wild plants are harvested for commercial use, populations that can't support such aggressive collection are often quickly destroyed. The loss of American ginseng from much of its native range is a cautionary example of how economic incentives for wild harvest can reduce a once plentiful species to a tiny fraction of its former numbers.

Perhaps, though, invasive species offer an exception. Because of their bad habit of forming dense stands or spreading over large areas, a useful amount of a plant can often be quickly collected in one place. In problem areas, there is a lot of material; nobody minds if it gets ripped up, and if a really intensive harvest completely wipes out a local population, that's not "environmental damage" but a bonus. Some "invasivores," who seek to control invasive species by hunting or gathering them for food, like to quip: "If you can't beat 'em, eat 'em!" Websites tell you how to catch or collect and cook them. Many invasive species are as nutritious and palatable as scarcer native species, and well worth the effort.

But there are many things that can be done with plants besides eating them. The purpose of this book is to encourage people to envision a variety of possible uses for the invasive plants of their region. Ideally, for the most useful invasive plants, harvest might become intensive enough to eradicate local populations altogether, or at least keep them under control without extra expense and use of poisons. Since few, if any, wild plants in their native habitats can stand up indefinitely to the onslaught of harvest for a global consumer market, it seems reasonable to suspect that aggressive harvest of invasives could have some effect on their numbers. Here, authors of chapters on a variety of plant uses describe some examples that may inspire readers' creativity.

Chapters 1 through 3 focus on encounters with individual plants that are, in different contexts, both invasive in the United States and culturally valued for culinary and other uses: kudzu (Katie Carter King, Chapter 1), wild taro (Peter J. Matthews, Mohammad Anwar Hossain, and Van Dzu Nguyen, Chapter 2), and Japanese knotweed or itadori (Tusha Yakovleva, Chapter 3). These authors suggest that a re-orientation of our attitudes toward such plants may bring rewards. Many other species might merit similar treatments!

Chapters 4 and 5 provide broad overviews of the subject of invasivory, or eating invasive species, both animals and plants, from different perspectives. The authors of Chapter 4, Alana N. Seaman and Alexia Franzidis, seek to spread enthusiasm for invasivory. In Chapter 5, Sara E. Kuebbing, Joshua Ulan Galperin, and Martin A. Nuñez present a more pessimistic and cautionary view (centered mostly on experience with animal species, but worthy of consideration for plants as well) and highlight legal impediments to uses of wild products in the U.S.

In Chapter 6, Kathy Voth explains how many weeds, including some that seem quite unpalatable to us, can be used as free livestock forage or fodder. Chapters 7 and 8, by Thomas Avery Garran and Wendy L. Applequist respectively, enumerate some invasive plants that are valued in Chinese and Western herbal medicine.

Chapter 9, by Franziska Eller, reviews uses of common reed, which is one of a few invasive grasses that have potential for use as biomass or biofuel for alternative energy production. Finally, Chapters 10 through 13 explore various artisanal or artistic uses for plant materials derived from invasive species, including uses in experimental architectural structures (Katie MacDonald and Kyle Schumann, Chapter 10), basketry (Katie Grove, Chapter 11), dyes (Theresa Hornstein, Chapter 12), and handmade paper (James Ojascastro, Chapter 13).

❖ ❖ ❖

Some words of caution are necessary. Any activity undertaken in nature will be controversial to some people, and collection of alien invasive species is no exception. In fact, there are particular issues, involving law, safety, and ecology, that should be taken into account by anyone considering use of an invasive species.

First, the law. As Kuebbing et al. note in Chapter 5, in addition to the normal legal restrictions on collecting plants from public lands or private property without permission, there are specific regulations regarding the harvest, transport, and occasionally even possession of certain particularly feared invasive species. These regulations are well-intentioned and mostly reasonable, meant to prevent the serious harm that can be done by improper practice. Accidentally dropping a single fruit or seed (or even stem fragments, for a couple of species) from a sack could disperse an invasive plant to new habitats, such as your yard.

As another example, this book doesn't include a chapter on wood uses, but it should be obvious that most invasive trees could be used to make wooden items or as firewood, depending upon their wood characteristics. Long-distance transport of firewood, though variably defined, is strictly regulated or prohibited almost everywhere, and for very good reason: invasive insects, which damage plants and carry exotic diseases, can be hidden under the bark and transported to a place where they can attack previously uninfested forests. Following such regulations carefully can keep some of our own native trees alive for future generations.

Laws and regulations vary greatly among states, and we cannot guess how regulations or their enforcement might change in the coming decades. Therefore, this book cannot and does not provide legal advice. People who skirt the letter of the law in private to provide for their own subsistence usually face little risk of prosecution. However, in the modern era, if one wants to use a plant commercially it is essential that legal obstacles be fully investigated in advance and any required permits be obtained. Then, additional federal, state, and perhaps local regulations constrain the manufacture or preparation and sale of many types of goods, especially but certainly not limited to products for consumption (foods and herbal dietary supplements). The would-be businessperson must be careful to identify all regulations in force related to a desired economic activity.

Second, if you're interested in either commercial or personal uses of invasive species, you need to be alert to safety hazards. Of course, in collecting any wild plant for consumption, you should be very sure you recognize the plant and could tell it apart from any unsafe look-alikes. To use it medicinally, you also need to know the traditional dose form and dosage and if there are any expected side effects or contraindications. Please consult authoritative references for such details.

Plants on badly contaminated land can accumulate toxic levels of heavy metals or industrial pollutants. Plants along roadsides, power lines, railways, or farm fields, in addition to being often at risk for such contamination, could have been recently sprayed with herbicides whose results are not yet evident. Even outside those areas, land managers usually want invasive species dead, and may go out of their way to poison them specifically. Therefore, before collecting anything to be consumed in any form by humans or animals, you should be confident that you know how the land it's growing on has been treated, especially recently.

None of that, of course, is intended as a blanket discouragement of such uses. Humans have been eating wild plants for millions of years. Done sensibly and knowledgeably, it is certainly not more dangerous to health than living on ultra-processed food from the supermarket.

Finally, there are ecological issues. Natural resources can be harvested sustainably (e.g., taking each year no more ginseng plants than will be added by growth from seeds) or unsustainably (taking everything one can find, so that the population soon goes extinct). The latter treats a renewable resource as if it were a nonrenewable resource, such as an oil well. Foragers, ethical wildcrafters, and practitioners of sustainable forestry learn not to do that. They learn principles to keep the plants alive for the next generation: Take only a small fraction of the plants you see. Pick scattered individuals from the middle of the patch rather than pulling up all of the plants at one side. Don't pull the whole plant if you only need the leaves. Leave parts of the root or rhizome behind. Harvest after fruit set and help to disperse the fruits. Beyond these common rules of thumb, they may study the specific biology of the species and landscapes they're working with to understand better what kind of harvesting will be sustainable in those particular circumstances.

Some wild plants actually benefit from moderate harvesting by such traditional methods, which thin out larger plants and make space for new plants, so that harvesting may not decrease population numbers at all (e.g., Reid, 2005; Huai et al., 2013; Poudeyal et al., 2019; see also the excellent and thought-provoking book *Braiding Sweetgrass*; Kimmerer, 2013). You do not want to do that to your local invasive species, which will take full advantage of the assistance! If you hope to control or even eradicate a population of an invasive species through harvesting, you must do the exact opposite of what you were taught was ethical behavior: Start at one edge and chew into the population, taking every individual in sight. Try to yank herbaceous plants up by the roots even if you won't use the roots. Harvest when the plant is young, if appropriate, and avoid letting it go to seed. Reproductive parts that aren't being collected for use should be carried away in trash bags to be discarded, if at all possible, rather than being abandoned on the ground where they could sprout.

Some scientists worry that if the public learns to use invasive species, it will result in increased spread of those species for either of two reasons: accidental promotion of growth by inadequately destructive harvesting, or deliberate spread of a species by people who are making money from it, so want it to continue to thrive. (See Chapter 5 for a thorough discussion of this perspective.) The former concern can likely be avoided, or at least minimized, by being deliberately destructive, harvesting the same area repeatedly, carefully observing the results over time to see if they are consistent with your intentions, and modifying your practices as needed. People living next to an invasive population might argue that the plants are already spreading successfully now, even though nobody is harvesting them, so they might as well provide the humans in their vicinity with benefits as well as costs.

As for the latter concern, only a change in attitude toward the products of nature, and what duties we owe in return for them, can truly avoid that risk. Sometimes "invasive" species are the only ones that can thrive on a piece of land because it's been so badly degraded by humans. In those cases, most people would think that it's better to have the invasive plant than *no* plants to protect or build the soil. Otherwise, most of us would agree that it's better to get rid of that kudzu patch.

Do not, therefore, start making and selling a kudzu product with the plan that you'll be doing that for the rest of your life; your purpose would be corrupted from the beginning. Think of the population as a nonrenewable resource from the moment you start to harvest it, even though that's not really true, and you'll envision its eventual "depletion" as being only natural. And if someday you actually manage to kill all the kudzu in your neighborhood? Then, you may not be able to keep working with kudzu—which you will have anticipated well in advance—but you can trust that the plants that will spring up in its place, though they may not be a handy monoculture, will have valuable uses of their own. Any healthy forest or grassland supplies us with many gifts, if we can learn to see them. The message of this book is certainly not that invasive species offer us more, but that even lands we have damaged do still have gifts to offer.

Literature Cited

Fantle-Lepczyk, J. E., P. J. Haubrock, A. M. Kramer, R. N. Cuthbert, A. J. Turbelin, R. Crystal-Ornelas, C. Diagne & F. Courchamp. 2022. "Economic costs of biological invasions in the United States." *Science of the Total Environment* 806: 151318. https://doi.org/10.1016/j.scitotenv.2021.151318

Huai, H., G. Wen, W. Xu & W. Bai. 2013. "Effects of commercial harvesting on population characteristics and rhizome yield of *Anemone altaica*." *Economic Botany* 67: 41–50.

Kimmerer, R. W. 2013. *Braiding Sweetgrass. Indigenous Wisdom, Scientific Knowledge and the Teachings of Plants*. Milkweed Editions, Minneapolis.

Poudeyal, M. R., H. Meilby, B. B. Shrestha & S. K. Ghimire. 2019. "Harvest effects on density and biomass of *Neopicrorhiza scrophulariiflora* vary along environmental gradients in the Nepalese Himalayas." *Ecology and Evolution* 9: 7726–7740.

Reid, L. A. 2005. *The Effects of Traditional Harvesting Practices on Restored Sweetgrass Populations*. M.S. Thesis, State University of New York, Syracuse.

CHAPTER 1

When Kudzu Ate the South, Why Didn't the South Eat Kudzu?

Katie Carter King

As a child raised to fear snakes, I knew better than to go near the kudzu. It licked at the edges of the cul-de-sac where I played with the neighborhood kids; it scaled the power lines that rose above the forest behind my home. I grew up unafraid of the woods, never concerned with a tick or a newt. But kudzu? That was something different. Something invasive, something unrelenting.

These primordial fears are top of mind as I scale a steep, unmarked drive in rural western North Carolina. Dead leaves crunch underfoot as I realize I've steeled myself to meet a forest of green monsters, lithe and writhing with cottonmouths, a vision of kudzu in late summer. But now, in mid-December, all that surrounds me are dormant vines braided into a loose net over the hillside.

By the time Lauren Bacchus, the executive director of a local nonprofit called Kudzu Culture, joins me, I am no longer nervous.

"You know," I say, gesturing wildly, "this is . . ." She finishes my thought before I do.

"Not scary?"

A version of this piece first appeared in *Gravy*, a quarterly journal from the Southern Foodways Alliance (No. 85, Fall 2022).

"Yeah," I say. "Not scary at all."

Lauren—or LB, as she prefers to be called—has had this conversation many times before. A 34-year-old North Carolina native, she's studied the attributes of this objectively useful plant. She begins to rattle them off for me. Erosion control, of course: the reason the Civilian Conservation Corps aggressively planted kudzu throughout the South in the 1930s. The roots have nitrogen-fixing qualities that replenish overused soil. Fibers from the spindly vines can be cured into hay, fed to cattle, or woven into delicate yet hardy textiles. In East Asia, where kudzu originated, as well as in holistic-health circles here in the United States, the root is understood to have myriad medicinal properties, from easing hangovers to healing snake bites.

In China, Korea, and Japan, foragers have understood these qualities for centuries, if not millennia. Community elders passed down their knowledge of how best to harness the legume's various functionalities, and, in turn, regular harvests in both summer and winter kept the perennial vine's growth contained. *Seikatsu roku* (Okura, 1828), a 19th-century Japanese monograph on kudzu, celebrates the plant as a "useful thing . . . in useless places." But perhaps the most useful application of kudzu is the one most underexplored in the West: the production of kudzu starch.

Kudzu powder, known as *kuzu-ko* in Japanese and usually shortened to "kuzu" in English, has been commercially manufactured in Japan since at least the early 17th century. An adaptable gelling agent, kuzu is traditionally used to thicken sauces and add body to custard-like desserts, in place of similar powders such as arrowroot, cornstarch, or gelatin. Many *wagashi*, traditional Japanese confections, rely on kudzu powder, like *kuzu-dama*, a red-bean cake coated with the starch. It appears in savory preparations as well, such as in *goma dofu*, a gelatinous, tofu-like block flavored with sesame; and *ankake*, an unctuous sauce poured over stir-fries.

Although kudzu grows across varied climates and terrains, certain conditions create the best kuzu. The roots are at their starchiest in early winter; after the greenery collapses, the plant draws nutrients back into its root network to survive a season of dormancy. The colder the winter, the more starch each tuber produces. According to *The Book of Kudzu* (Shurtleff & Aoyagi, 1998), the definitive English-language guide to kudzu, production is optimized when plants are grown with ample sunlight around a latitude of 34 degrees north, typically on steep mountains that climb to about 2,500 feet in elevation.

Standing on this peak of sleeping kudzu in Madison County, North Carolina, I check my phone. We're roughly 2,600 feet above sea level, looking down over small farmsteads and fields. Some are clearly in use, while others appear overgrown. Our latitude is 35 degrees north. This high in the Blue Ridge Mountains, there's little topographically induced shade. And on this crisp, clear day, with the trees mere skeletons of themselves, it seems as though the sunshine could find you in the deepest part of the forest.

"Around Nara and Yoshino, where the best kudzu starch comes from," LB says with a grin, "it looks just like this."

❖ ❖ ❖

Kudzu's overgrown American tenure has been well-documented (Fig. 1). First imported as an ornamental in the 1870s, wealthy planters up and down the East Coast prized the vine for its tropical ambiance. They trained the dense foliage to climb porch railings and fence posts to foster both shade and privacy. Early adopters, such as Quaker conservationist Charles Pleas, immediately noted the vine's rapid growth. By 1917, researchers at the Agricultural Experiment Station in Auburn, Alabama, were studying kudzu's usefulness as cattle fodder. Although farmers were initially skeptical of the zealous crop, by the early 1930s, the South's malnourished soil needed large-scale intervention. As ethnobotanist and retired North Carolina Cooperative Extension agent David Cozzo later explained to me, in much of the South—such as in the Georgia Piedmont—there was more than 11 inches of rich topsoil covering the ground when commercial agriculture began to spread in the 18th and 19th centuries. Southern planters' commitment to monocropping, whether cotton or tobacco, left the earth stripped of nutrients and vulnerable to erosion. "All this topsoil, by the 1950s, was in the rivers. It washed off the land and left nothing but red clay," he says. "If it wasn't for exotic, invasive species, there wouldn't be any plants left in the Piedmont."

In 1935, the Secretary of Agriculture formed the Soil Conservation Service (SCS), an agency dedicated to combating widespread erosion. Under SCS guidance, federal workers began planting imported kudzu shoots on thousands of acres of demonstration plots from Virginia to Alabama. Highway departments sowed kudzu on rights-of-way, while railroad companies embedded young vines into crumbling train tracks. The SCS even paid farmers $8 an acre to plant it on their own land, shipping seedlings directly to growers' doors. By 1939, kudzu had taken root in every state in the American South.

Then, of course, the flowering vine adapted far too well. Following World War II, agriculture became increasingly mechanized, requiring equipment

FIGURE 1-1. Kudzu in Madison County, North Carolina. (Photo by Maddy Alewine.)

that most homesteads were unable to afford. Automation, in turn, led to widespread job loss, as farms no longer required the same workforce they had only a few years prior. As folks fled to the cities in search of new opportunities, kudzu crept into the rural lands left behind. And without any other long-term intention for its use beyond erosion control, kudzu spread unchecked. (Natural predators that often attack new-growth kudzu, such as hungry deer and the aptly named kudzu bug, were not yet plentiful enough in North America to make much difference.) Public and scientific favor quickly turned against the vine, and in 1953 the USDA quietly removed kudzu from its list of promoted cover crops. Alabama outlawed the possession of kudzu seeds in the 1960s, and states from Florida to Massachusetts followed suit. Congress added kudzu to the Federal Noxious Weed List in 1997, and in 1999 *Time* declared its widespread cultivation one of "The 100 Worst Ideas of the Century" (August et al., 1999).

❖ ❖ ❖

As LB and I climb back down the steep trail, she stops to point out each individual part of the legume: the root crowns embedded in the dirt, smelling herbal and pungent; the blanket of brittle, intricate vines atop them; the seed pods crowning the growth, the only visible sign of the vine's blossoms remaining in December. "In one generation," she says, "we lost the little bit of cultural value it had."

In the modern American consciousness, kudzu arrived on our shores with malintent. Georgia lore said that if you didn't close your windows at night, kudzu's slender fingers would creep in while you slept. These anxieties are so ubiquitous, they've become a punchline: Southern humorist Lewis Grizzard once joked that the leaves could cover an entire house in a single evening, before musing, "Those who try to eat their way out of kudzu quickly have their innards entangled in the vine because no matter how much you chew it, the blamed stuff just keeps on growing" (Grizzard, 2007). While scientists have repeatedly found positive attributes hidden beneath the mass of tri-tipped leaves—from its availability as a biofuel to its usefulness in regulating blood pressure levels—these findings have done little to change kudzu's weedy reputation.

But where Americans tend to revile kudzu, others see great potential. Japanese kuzu producers have historically foraged, rather than cultivated, kudzu. But in 1990, Inoue Tengyokudo, a century-old kuzu producer based in Nara, Japan, purchased 165 acres of land in Lee County, Alabama.

Less than 30 miles from where Auburn's Agricultural Extension was then researching how to loosen the vine's grip on the region, the conglomerate planned to open the world's first kudzu farm. But harvesting kudzu roots is labor-intensive and best done by hand, in part because the roots tend to become tangled with their underground neighbors. The Tengyokudo experiment proved a swift failure. "They grew some very good kudzu," Auburn University agronomist John Everest told the Athens alt-weekly *Flagpole* (Britt, 1999). "The deer wiped the company out."

For 150 years, kudzu has remained anchored in Southern soil—an ornament, a miracle, a pest. It has been overplanted and underutilized, loved and loathed. But, throughout its American tenure, kudzu has never been what LB and the rest of Kudzu Culture understand it to be: an opportunity to find balance, both ecologically and economically.

Kudzu Culture began in 2011 as a loosely organized group of neighbors and friends with the mission, as they playfully put it, of "eating the vine that ate the South." Initially led by Zev Friedman and Justin Holt, two western North Carolina–based permaculture educators, the pair were drawn together by their shared interest in designing ecologically diverse and resilient communities. Friedman and Holt would host biannual, three-day events called Kudzu Camps, where they invited community members to learn the intricacies of processing the different parts of the plant. In winter, they dug out the roots for starch extraction; in summer, they harvested everything from the vines to the flowers for a variety of craft and culinary applications, such as baskets, jellies, and woven garments. LB, a fiber artist long interested in creating textiles from kudzu vines, connected with the group after she moved to the area from her hometown of Davidson, North Carolina. In the years since, the trio has been brainstorming how best to provide comprehensive, year-round education and training resources for local people to work with the plant, rather than against it. "Combatting invasive species is a place where there's a lot of resources and energy and attention being directed," Holt explained. "There's been kind of a knee-jerk reaction to invasive species in general, but people are open to it."

"We're trying to increase the awareness of their value and help people develop a different kind of relationship, one that we think is more life giving," he continued.

However, making kuzu is far from easy work. Extraction is a delicate and labor-intensive event, as the starch-rich roots can be more than six feet in length, seven inches in diameter, and weigh more than 400 pounds. Each root must be individually dug out by hand: any cuts made directly into the flesh enables starch fermentation, significantly lowering the quality of the final product. But, if properly produced, kuzu can be valuable—Cozzo says he last saw it retailing for $36 a pound in an Asheville health-food store before the pandemic, while LB recently found it selling for more than $60 a pound online—in part because it is incredibly expensive to make, especially on a commercial scale.

Whether on a Japanese mountainside or a western North Carolina farmstead, kuzu production is generally the same: After careful unearthing, processors crush each root, creating an almost pulled pork–like texture, before submerging them in water and sieving the slurry to remove any large fibers. The remaining thick, milky solution stands until the heavy starch sinks to the bottom, while the impurities and other unwanted particles rise to the top. Once this cloudy liquid is removed, the starch is re-washed and allowed to settle again. This process is repeated continuously until only pure white starch remains. Cold water and cold weather ensure that the starch does not ferment throughout this process, which in a traditional, family-owned Japanese shop can take up to 90 days.

Kudzu Culture first experimented with refining starch using whatever equipment they had access to, from simple buckets to a repurposed old washing machine. ("The centrifuge helps to separate that heavier matter from the liquid," LB explains.) But to sell kuzu commercially, they'd need USDA certification. To avoid the financial investment of purchasing industrial, food-safe shredding and extraction machines, the team hoped to find a "white label" producer—a pre-existing potato- or cornstarch manufacturer that would process the foraged kudzu roots in exchange for a cut of profits. But, as LB found, all American-based starch processing facilities were owned by one of a handful of big corporations, and the trio was unwilling to pursue a contract with a multinational conglomerate. Apart from independent artisans, "domestically, there's no kudzu starch producer or kudzu fiber producer other than us right now," she says. While a handful of family-owned kuzu producers still exist in Japan, in the last half-century the global forces of industrialization and agricultural consolidation have also affected kuzu production in Japan. Only a fraction of the kuzu businesses that existed before World War II are still in operation, and younger generations do not have the same understanding of starch production their forebears did. With fewer foragers to keep root growth in check, kudzu has started to exhibit invasive growth patterns in some areas—meaning it's begun to eat Japan too.

Although the number of kuzu producers has fallen greatly in the last 50 years, globally, demand has continued to rise. Health food stores across the

United States started carrying kuzu in the 1960s, as holistically minded consumers began to seek out more traditional forms of medicine. Many were inspired by the healing properties the starch was supposed to store. (Cozzo, who is originally from New York, says he encountered the powdered root in macrobiotic food stores more than a decade before he ever glimpsed a thicket of kudzu.) The influx of East Asian immigrants and refugees in the late 20th century increased the demand for kuzu, which can be found today in many international grocery stores.

In recent years, a handful of restaurateurs across the South have begun to grasp the culinary possibilities of the invasive plant. Kudzu Culture previously sold kuzu to Sean Brock in an experimental capacity. Brock, the Nashville-based owner of Audrey, is one of several chefs interested in the possibilities of a locally produced, Southern starch. Other culinarians infuse teas with kudzu flowers, long clusters of grape-scented petals that Southern women have traditionally turned into jelly. Asheville-based Shanti Elixirs already brews kudzu jun, a fermented tea similar to kombucha, using Kudzu Culture's foraged goods. Shanti Elixirs currently sells two seasonal products: a kudzu-dandelion-chicory drink in colder months, and a summertime brew flavored with Asian pears, ginger, and kudzu blossoms. Dehydrated kudzu can be turned into a simple, yet flavor-rich tisane on its own, with notes of sweet vanilla tinged with an edge of burdock-like bitterness.

Kudzu Culture believes the legume has the potential to become a staple ingredient in restaurant kitchens across the South. Still, the group has had trouble securing funding over the years, largely because of kudzu's invasive reputation. But oft-cited numbers purporting that kudzu covers somewhere between seven and nine million acres across the United States have come under recent scrutiny. Assessments from the United States Forest Service estimate the total figure at less than a million acres. (By comparison, privet, brought to America in the 1700s as an ornamental hedge, now grows on more than three million acres across the country.) Kudzu Culture finally received their first grant early in 2022 from the Educational Foundation of America, a $20,000 investment toward creating a community root and starch processing space that will be located in Blue Ridge Food Ventures, a commercial processing and food-packing facility near Asheville.

While they work to commercialize their kuzu operation, Kudzu Culture is planning farmer-harvester trainings that teach interested folks how to appropriately excavate, clean, and process the roots for starch. "We're trying to build the skillset of the people who are digging roots, so they know precisely what we're looking for and what to avoid and how best to do it,"

Holt said. If an individual wants to harvest on their own land, the group will perform a site visit to certify that the kudzu is grown in uncontaminated soil. Once harvested, Kudzu Culture will buy the roots from people in the community as part of their "Root Buy-Back" program. Currently, they're paying $3 a pound. They purchased about 150 pounds of roots from local residents in the first year of the initiative.

This sort of work has a historic precedent in western North Carolina. Cozzo studied the root digger economy, where locals have long supplemented their income with wild foraging. "That type of economy thrives in a low-income area," he says. "While you're out hunting, you go gather some ginseng. Then, at the end of the season, you'd sell your ginseng and that'd pay your property taxes for the year." Whole families and sometimes entire towns would gather ginseng and other medicinal roots, similar to how multiple generations would harvest kudzu together in Japan.

Kudzu Culture doesn't expect this work to be a quick fix or a wholesale salve to the larger historic and systemic problems rural western North Carolina communities are up against, such as land degradation and limited economic opportunities. Rather, it's a generations-long quest to replicate the balanced ecological systems that sustained communities both here and in Japan for centuries. In the short term, the group sees it as a way to engage with longtime residents and provide direct access to capital, all while pushing Southerners to engage with their natural surroundings differently—no matter if the flora is invasive or native, abundant or ephemeral.

"Abundant species" is a term LB relies on, as she finds it contains more possibility and less prejudice than a word like "invasive." Besides, as she tells me, "The vilification of abundant plants at this stage of the Anthropocene"—a reference to the modern environmental era defined by human presence—"is missing the point." In other words, kudzu isn't going anywhere. Rather than work to eradicate it, why not try to understand the latent possibilities it holds?

"That's the important thing not to miss," LB says as Japanese knotweed, another imported species, whistles in the December breeze. "We have opportunities to heal through humility." She's speaking of the humility to reexamine misunderstood history, to confront long-held prejudices, and to look to the future for answers, rather than to an idealized past. As people across the South, and across the world, grapple with the inequalities wrought by land misuse and rising temperatures, LB believes this kind of reframing will become key to forming strong, adaptable communities. But for now, she'll settle for kudzu.

Literature Cited

August, M., H. Barovick, M. Derrow, T. Gray, D. S. Levy, L. Lofaro, D. Spitz, J. Stein & C. Taylor. 1999. "The 100 worst ideas of the century." *Time*, 14 June 1999, 153(23): 37.

Britt, J. 1999. "Why can't anybody do anything with kudzu?" *Flagpole*, 22 September 1999, 13(38): 10–11.

Grizzard, L. 2007. "Put some South in yo' mouth: Lifespan in Georgia." *Y'all*, December 2006/January 2007: 67.

Okura, N. 1828. *Seikatsu roku*. Bunshodo Fujii Uhei. [In Japanese.]

Shurtleff, W. & A. Aoyagi. 1998. *The Book of Kudzu*. Soyinfo Center, Lafayette, California.

CHAPTER 2

Welcome to the Green Food Zone: Wild Taro (*Colocasia esculenta*)

Peter J. Matthews, Mohammad Anwar Hossain & Van Dzu Nguyen

Taro, *Colocasia esculenta* (L.) Schott, is a global cultivated food crop, familiar to billions of people in Asia, Africa, Oceania, and the Americas (Fig. 2-1; Matthews & Ghanem, 2021). Although widely known as both a starchy root crop and a green vegetable, the specific uses of taro vary greatly according to the varieties cultivated and cultural preferences (Matthews, 2010). The cooking knowledge that is passed down across generations in families and communities is mostly related to cultivated taro varieties (cultivars). Some of this knowledge has been transferred to cookbooks and websites, in English and other languages, but knowledge of how to recognize, process, and cook edible wild taros (i.e., wild forms of *C. esculenta*) is highly localized and has been studied even less than the plants themselves.

The natural range of taro (with many wild breeding populations) is thought to be from India to China, Indonesia, Papua New Guinea, and northern Australia, and cultivated forms of taro originated somewhere within this natural range (Matthews, 2014; Matthews et al., 2017; Ahmed et al., 2020). Wild taros grow inside and outside the natural range and come in many different flavors, so to speak: from the unspeakably itchy (they can cause throat swelling and toxic shock) to very good tasting when harvested and prepared by an experienced cook. The quality of any particular wild taro as a food depends

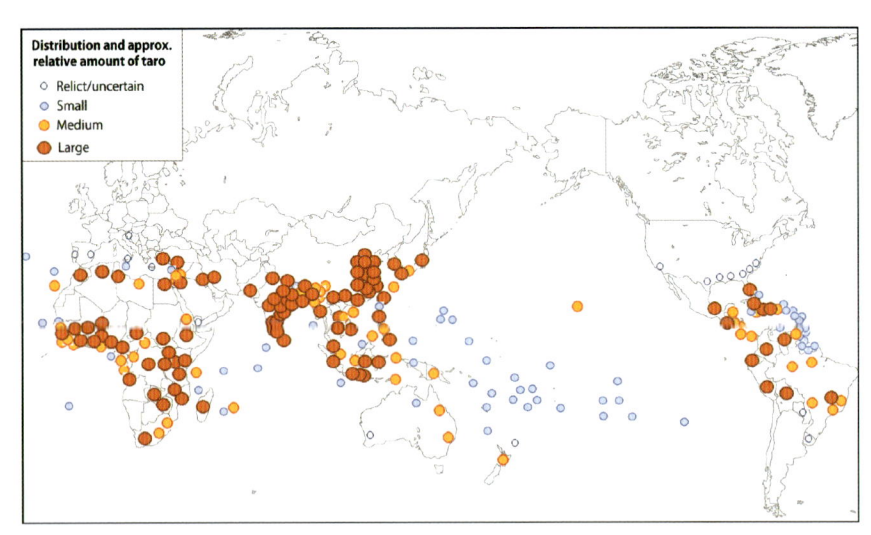

FIGURE 2-1. World distribution of taro under cultivation, with size of circles representing relative amount of production (Matthews & Ghanem, 2021).

on variety, age of the part eaten, growing conditions, and knowledge of the cook. Cooking methods that work with a wild taro in one area might not work in another area. In this chapter, the phrase "wild taros" refers to the diverse wild forms of *Colocasia esculenta*. Wild taros include natural populations that range widely in Asia and the western Pacific, and populations established by human activities in both these regions and far beyond.

First Steps

Here, only general advice and principles are given for cooking taro. The two most important points are: (i) ask people living nearby for any advice they can offer based on direct experience with the plants on hand, and (ii) carefully test cooking methods whenever using plants from an unfamiliar source. Being careful in these ways is needed not just when using wild taros, but also when using unfamiliar cultivated varieties from a garden or market.

For those who are completely unfamiliar with cooking taro, it may be best to begin with tubers (starchy storage organs that are commonly called corms) or leaves from a vegetable market, since these will certainly be edible after common basic steps are taken. If the same basic steps do not work for a particular wild taro, then additional steps may be needed to make it edible.

In most large cities around the world, in Europe, the U.S.A., and Asia, taro is sold under different names according to where sellers obtained their stock, or according to the names familiar to those particular sellers. In Europe and the U.S.A., the sellers are often immigrants from Asia, Africa, or the Caribbean. Taros from Cyprus and the Atlantic Islands (especially Canary Islands and Azores) are also found in certain European markets. Regular customers will take the freshest products quickly, so what remains in the bottom of a nearly empty box is likely to be in poor condition.

When buying corms that are still in the skin, look at places where cuts have been made (top and bottom), or where corms have been separated from each other (scars). These are the places where rotting can easily begin. The corm may be visibly shrunken in those areas, or may be soft under the dry skin. On a large corm, it is often possible to recover an edible portion by cutting away any soft or discolored parts, but the corm as a whole may not be fresh and good tasting. Discoloration is first apparent when otherwise white fibers in the corm flesh become brown. Rotting appears to spread first in the fibrous vascular (liquid transporting) tissues. A fresh, starchy corm will snap apart when partly cut then levered with a knife blade. When not fresh, with less starch and more obvious fiber, the tissue has more spring or does not snap so easily. Fresh taro is a healthy food that satisfies our appetite for longer than sugary or highly processed foods. A good visual test, for corms from the wild or market, is to cut across the corm to expose a cross section, then press the knife edge into the surface so that liquid is squeezed onto the blade. If this liquid is clear, then you are looking at an old section of corm with a lot of fiber and not much starch. It may be edible, but will not become soft and tasty when cooked. If the liquid expressed is opaque or milky, then you are looking at a high-quality section of the corm. Buying older, cheaper corms may not be cost effective if most of each corm has to be discarded.

Side-corms are the vegetative or clonal offsets or "children" of a mother corm (Fig. 2-2) and are often fresh and starchy from top (the shoot end) to base (where it is attached to the mother). Depending on the variety and location, the larger mother corm can take up to several months or more to grow. Soil fertility, water supply, sunshine or shade, and temperature all affect the speed of growth, but good results can be obtained under many different conditions. What is "good" also depends on what cooks consider acceptable for the particular dishes they want to prepare. Corms of different size, shape, and internal quality can all be used in different ways: new users should experiment to find what suits their own taste. Corms vary in density, hardness or softness, stickiness or dryness, flavor, and so on.

A mother corm is usually starchy throughout and good to eat after a single growing season. If left in the ground for more than one season, there may be obvious differences in quality between the younger upper and older lower parts. Plants that mature quickly or are left long enough in the ground to flower use stored starch to support the development of inflorescences, leading to a harder texture after cooking (which some people prefer). The best time of year to harvest wild plants varies according to the part consumed, variety, and climate but is always autumn or winter for corms in temperate regions, after new leaves start becoming smaller, or after the plants have stopped growing completely.

According to cost, taste and texture preferences, and need, cooks may discard some of the less starchy, lower part of a mother corm, even if it is edible and not discolored or rotten. Corms are easily stored and transported, locally and internationally, but require cleaning and careful observation to avoid transmitting soil, nematodes, viruses, fungal pathogens, and insect pests. A small amount of dry soil on fresh locally produced corms is not usually a problem and indicates that there has been less handling (to clean corms) and therefore less chance of damage. Information about taro pests and diseases can be found in many publications.

The green parts of taro that can be eaten are the blade, petiole, whole leaf (blade plus petiole), and young stolon (Figs. 2-2, 2-3). These are usually only sold in the area or country of production. Dried leaves (blades and petioles) can be stored for long periods and are traded internationally. Old leaves are never eaten, and the edibility of blades, petioles, stolons, child corms, and mother corms all vary according to the variety and cooking methods used.

Taro is able to propagate itself and spread naturally using vegetative shoots (either child corms or stolons) that grow near the parent or break off and disperse in fresh water (Fig. 2-4). New plants grow from corms or skin peelings with buds, or from stolon nodes (clonal, vegetative propagation). In tropical Asia and the western Pacific, it also produces fruiting heads with sweet, many-seeded berries (Fig. 2-3B) that are attractive for birds and other animals (Matthews & Naing, 2005). Outside this region, reports of flowering, fruiting, and seed production are rare, except in Hawai'i, where close observation of the crop in a subtropical environment has led to many reports, though insect pollinators for taro have not been reported there. In tropical Asia and the western Pacific, most pollination is by small flies (*Colocasiomyia* spp.) that are specialists for taro.

GREEN FOOD ZONE: WILD TARO 23

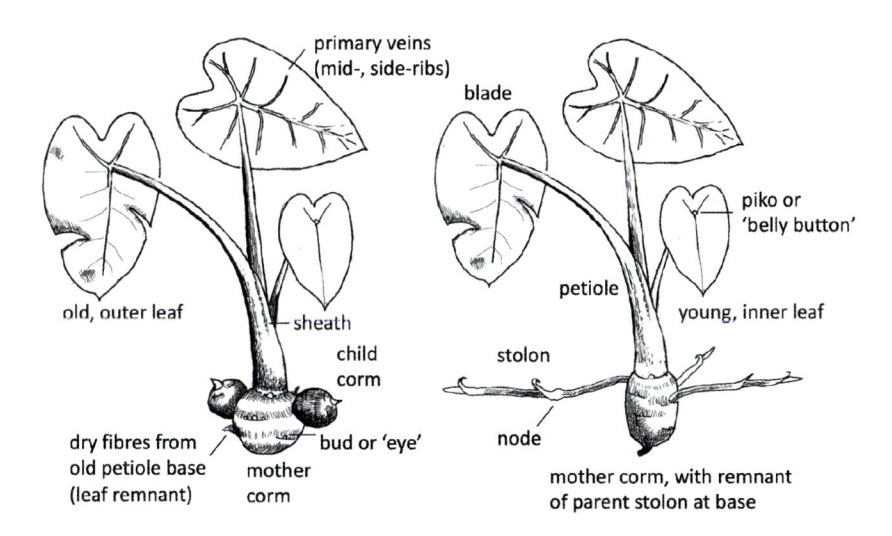

FIGURE 2-2. The two main morphological varieties of taro, with child corms (left) or stolons (right), and other plant parts labeled. There are many different forms within each variety, and many intermediate, possibly hybrid forms. The Hawaiian term *piko* (meaning belly button) is used here as no single English word exists to identify the junction of petiole and blade viewed from above. As with all parts of taro, the piko color (IPGRI, 1999) can vary and be used to distinguish different varieties.

FIGURE 2-3. Wild taro, *Colocasia esculenta* var. *aquatilis*, in Vietnam. (**A**) Whole plant with edible stolons, including snapped segments of size used for eating. (**B**) Left, near-mature fruiting head with many green berries containing mature seeds. Right, mature orange-red berries. (**C**) Left, immature fruiting head with lower spathe removed. Right, spadix with full spathe removed. (Photos by P. J. Matthews, various locations, 2011–2017.)

FIGURE 2-4. Stolon and sprout of wild taro, *Colocasia esculenta* var. *aquatilis*, floating and tangled with other plant debris in a flooded lake after heavy rainfall. (Photos by P. J. Matthews, Ba Be, Vietnam, 2012.)

Into the Green Food Zone

As a food plant for people, vegetative parts of taro have been widely carried and planted, and usually—but not always—cultivated. In New Zealand (Matthews, 2014), Myanmar (Matthews & Naing, 2005), and other countries, taros have been deliberately planted in suitable environments without subsequent cultivation, in roadside ditches, wet gullies, on stream banks, and next to springs. In later years, the resulting wild clumps or patches can be harvested and replanted at the same time. Taro responds to being cut by putting out new side shoots and propagating itself. If harvesting is not frequent or continuous, then replanting is not needed, though it can improve spacing and yield.

Wild taros are part of a green food zone that exists around human settlements in many regions of the world. These are places where the ground is not cultivated, but vegetation is often cut back to keep roads, trails, canals, ponds, ditches, and the edges of fields and gardens clear for access or to keep views open. Such locations can be described as "modified" or "disturbed," and plants that occupy these edge habitats are often identified as "ruderal." Edge habitats exist between open, cleared land and various plant communities created by people such as the village forest shown in Figure 2-5B. The green food zone also extends to natural edge habitats such as the sides of streams, rivers, lakes, and waterfalls, where frequent flooding or overflow prevents tall vegetation (shrubs, trees) from growing and shading out smaller herbs. In tropical Asia and the Pacific, wild taro populations occupy a wide range of modified and natural edge habitats. The term "wild" is just a general way of distinguishing places that are not cultivated or otherwise closely controlled.

FIGURE 2-5. The green food zone and commensal wild taros. (**A**) Taro planted in the "Green Food Zone" at edge of a public road (Korea, 2013). (**B**) Taro wild at the edge of village forest and pasture; grazing animals such as the cow, water buffalo, or goat do not touch acrid taro plants (Philippines, 2013). (**C**) Taro in pond at edge of city, together with the South American water hyacinth (Bangladesh, 2019). (Photos by P. J. Matthews.)

In areas where wild taros and other edible wild plants are commonly used, commonsense rules of etiquette usually apply: only take from plants that are growing on public land or waterways, don't take more than you need, don't take from immediately in front of a private house where people are looking after and using a patch (or ask permission first), and replant loose or discarded shoots if you or others are likely to visit often in the future. Repeated harvesting of the same patch is said to improve eating quality (Teron, 2019), presumably by promoting new growth.

In 2013, in the warm temperate region of southern Korea, taro, corn, and other vegetables were found growing along a narrow strip of soil at the edge of a busy city road, with a large signboard announcing a "Green Food Zone" (Fig. 2-5A). This deliberate effort to establish food production in a public space illustrates how wild populations can be started by human action: if cultivated, weeded, and replanted, the strip will continue as a roadside garden; if not carefully maintained, the corn will disappear and wild herbs will invade, but the taro will likely persist with corms that are dormant in winter and sprout in spring. In northern New Zealand, another warm temperate region, there are many wild taro patches in ditches and stream banks along public roadsides or in open farmland, and these are often harvested by local communities (Matthews, 2014). These patches do not spread much because the plants mainly produce side-corms. The latter can separate from the parent and wash away in floods, but usually sprout while still attached or nearby after the parent has rotted. In Japan, cultivars produce many side-corms that vary in size (e.g., Fig. 2-6A) and discarded corms often sprout at the edges of fields, in ditches, or on streambanks. Figure 2-6B shows a side-corm sprouting on a streambank in India.

Many wild and cultivated varieties of taro in tropical countries grow in wetlands and have long-creeping stolons (also known as runners) (Figs. 2-2, 2-3, 2-7). Like side-corms, these are side shoots that sprout from buds inside the base of each leaf (axillary, lateral buds). Plants that produce side-corms generally do not produce long stolons, and vice versa. These are alternative forms of side-shoot, though some cultivars produce elongate side-corms that are attached by short stolons, or produce short stolons under certain conditions as well as side-corms. It is mainly plants with long stolons that have come to be regarded as "invasive" or "weedy" in areas where they are not regarded as useful wild plants. Reports of taro as invasive or weedy generally come from countries (or areas within countries) where the dominant culture is European in origin, including Australia, Spain, and the U.S.A. In modern European cooking traditions, taro is not familiar as a food plant.

FIGURE 2-6. Taro side-corms. (**A**) Dryland taro cultivar in Japan (*kinu-hikari*), whole plant with 10-cm scale; mother, child, and grandchild corms separated below; similar cultivars with many small side-corms are widespread in cooler montane or temperate regions of Himalaya, eastern Asia, and northern New Zealand. (**B**) Isolated side-corm sprouting on sandbank next to a boulder stream near Tura, northeastern India, presumably after dispersal in water. (Photos by P. J. Matthews, 2015.)

Remarkably, the oldest known collection of recipes in Europe, *Apicius de re Coquonaria*, includes recipes for taro under the name of *colocasia* (Vehling, 1977), which is an old Latin and Italian vernacular name for taro, and clearly related to the Greek name *kolokasi*. Taro is widely cultivated in the eastern Mediterranean today and is eaten in Turkey, Lebanon, Syria, Cyprus, and Egypt (Matthews, 2006; Grimaldi et al., 2018). It also spread across northern Africa and was introduced as a food plant to Spain, Portugal, the Canary Islands, and the Azores. In Italy, it survives not as an invasive plant, but as a treasured ornamental plant growing in water gardens and under fountains from Rome to Sicily. Many free-flowing rivers and streams in Italy have become seasonal rather than perennial, as water is diverted for agriculture and cities. Nineteenth century records indicate that taro previously grew as a naturalized, wild plant in southern Italy (Candolle, 1885). In the United States, taro has been introduced from many sources, as an imported crop, at different times in history, so there may be considerable diversity in the plants that now grow wild.

In 1905, Safford reported that: "Taro is imported into the United States from Canton [in China] and the Hawaiian Islands, and is sold in large quantities in the Chinese markets of San Francisco. It is successfully grown in southern California" (Safford, 1905). He also cited reports of experimental production of taro in Florida in the 1890s, but taro first reached the southern U.S. long before then. It may have first arrived as a crop from Africa together with Africans for whom it was a familiar and welcome food (Catesby, 1781; Carney & Rosomoff, 2009). Wilson (1960) provided what may be the best snapshot of the spread of cultivated and wild taros in the southern U.S.:

> Numerous cultivars occur, and several have been introduced into the United States, where the species has been grown in the lowlands of the Coastal Plain from South Carolina to Eastern Texas. *Colocasia esculenta* var. *aquatilis* Hasskarl has escaped cultivation and is an aggressive weed which forms large clones spreading vegetatively by slender, rapidly growing stolons. . . . It has been reported to be spreading in southern Louisiana and is also known from several localities from central to southern Florida growing along streams, marshes and roadsides.

This example of the spread of taro as a cultivated and then wild plant in the U.S.A. illustrates a process that has taken place in many countries with warm and wet climates, inside and outside the natural range of taro. In Southeast Asia, where the crop is thought to have been domesticated, taro has spread with people and also naturally as a wild species. This has led to a complex situation in which natural wild forms that are inedible (or difficult to process for eating) exist alongside edible wild forms of taro that were deliberately planted in wild habitats or spread naturally (spontaneously) from kitchen dump sites and gardens.

The natural range of taro (Asian mainland to Australia and New Guinea) is a warm and generally wet region that corresponds approximately to the Asian monsoon zone. Outside this range, the plant has been mainly introduced for cultivation as a food crop, most notably in Africa, the Mediterranean, and Oceania, during antiquity (Matthews, 2006, 2014; Matthews et al., 2017; Grimaldi et al., 2018). The use and spread of taro as an ornamental plant is mainly linked to modern urbanization and international trade, and ornamental taros are often derived from edible cultivars (please note that various wild species of *Colocasia* are also sold as ornamental plants and should certainly not be eaten; these are not regarded as "taro" in this

FIGURE 2-7. (**A–B**) Harvesting the young stolons of wild taro (*Colocasia esculenta* var. *aquatilis*) from edge of a lake in limestone hills. (**C–D**) Peeling stolons for use as green vegetable in various dishes, including a stir-fried dish with garlic, meat, and soy sauce. The gloves are optional and prevent itching of the hands. Young stolons are thicker, softer, and probably less acrid than older stolons. Peeling to remove the fibrous epidermis also helps to reduce acridity. (Photos by P. J. Matthews, Ninh Binh, northern Vietnam, 2017.)

chapter). In most countries where the plant has naturalized, outside the natural range, the wild populations are likely to be edible—even if corms are better for eating after cultivation in fertile soil with adequate water, in warm and sunny conditions. Wild taro varieties that are regarded as invasive weeds outside the natural range may in fact be elite eating varieties that were originally taken to distant countries because of their prized food qualities. Part of the process of becoming "naturalized" may be losing knowledge of the plants as good and useful, and knowledge of how they came to be present in a particular area. More study in the southern U.S. (for example) might reveal that some communities (descendants of African, Asian, Pacific, Mediterranean immigrants?) are connected to local wild taro populations, manage them, and recognize them as useful . . . as was found in northern Aotearoa/New Zealand in the 1980s when wild taros were surveyed in Maori community areas (Matthews, 2014).

In situations where knowledge is lacking or not shared by all people in a given area, the position of wild taros in the "green food zone" is problematic for three reasons. Firstly, not everyone is familiar with protocols for harvesting and managing wild taros. Secondly, taro (whether cultivated or wild) is an acrid, poisonous plant (see below), so for people who are not familiar with a particular wild taro variety, suitable methods for preparation and cooking need to be learned, with care and safe testing to avoid unpleasant and potentially dangerous experiences. Thirdly, taro might be introduced into new areas without care to select a non-invasive type (unable to produce seeds, or without stolons) or to manage the selected variety and prevent it from becoming invasive beyond the point of introduction.

When wild taros in a given area are recognized as valuable, and are treated as a public, open-access or community-owned resource, then shared understanding and practical rules are needed to manage and appreciate the plants. In countries as far apart as Bangladesh, New Zealand, Japan, the Philippines, Myanmar, and Vietnam, it is generally accepted that wild taros growing on public land can be harvested by anyone, but with care not to completely remove all plants or prevent them from growing back for others to enjoy in the future. In the case of taro, discarded or unwanted corms and stolons can easily grow again. In countries like New Zealand where growth is limited to just part of each year, and the plants eaten do not produce stolons, it may be important for people to replant unused corms or cut tops to ensure continuity of the food source.

In Bangladesh—where taro is a deeply appreciated crop, and where wild taros are often eaten—there is a terrifying metaphor: *kochu kata* means literally to cut taro (*kochu*) into small pieces, but it also means to "destroy ruthlessly" as a threat to human opponents or enemies. If we can love and let live, then the need to cut and destroy taro will diminish. There is no absolute need for hard or strict boundaries between "wild" and "cultivated," or "natural" and "modified." In reality they are impossible to maintain forever, and attempts to establish and keep boundaries can be counterproductive, resulting in the highly degraded ecosystems of modern industrial agriculture. Societies that coexist with animal and plant wildlife can enjoy high food productivity, high biodiversity, and greater resilience to changes in the social and physical environment. Such productivity is generally not measured or measurable in cash terms, and the uses of local wild foods such as taro do not appear in most statistics concerned with food production and trade.

The ability of taro to invade some regions may reflect larger problems of past forest clearance or broad-spectrum habitat destruction. And as a cultivated, flood-tolerant plant, taro may facilitate new movements by people into wet forest environments, resulting in further forest clearance. With its tall and large leaves, wild taro is highly visible as an introduced plant in deforested wetlands, or on open riverbanks (Everitt et al., 2007). It is not clear that the plants pose a threat to non-native organisms, though "visibility" can be a consequence of producing a continuous shade-forming canopy (Figs. 2-5C, 2-7A). The ecological consequences of invasion by taro in unmodified environments have not been studied, perhaps because so few unmodified environments still exist. Apparently natural (and genetically distinct) populations of taro in Australia's Queensland rainforest exist as highly constrained and sporadic local populations scattered widely but thinly in the landscape: they remained largely invisible to non-indigenous visitors until a special effort was made to find them (Matthews, 2014).

The present authors are familiar with the use of wild taro populations in Vietnam, where at least five wild *Colocasia* species are present (*C. esculenta, C. lihengiae, C. menglaensis, C. spongifolia,* and *C. yunnanensis*) (Matthews et al., 2022). Wild *Colocasia* species other than *C. esculenta* are not eaten, and not all wild populations of *C. esculenta* are eaten—some belong to a wild evolutionary lineage that is never cultivated (Ahmed et al., 2020), and rarely eaten (in northern Australia, wild taros in this lineage were previously eaten, but only after long processing [Matthews, 2014]). Wild, breeding populations of edible taro belong to a different evolutionary lineage (Ahmed et al., 2020) and are usually found growing commensally (with people) in modified wild habitats inside and around villages and fields, from southern to northern Vietnam. They flower, fruit, produce seed, and also spread spontaneously by stolons which disperse by extension across ground, or by breaking then transport in water (the stolons are brittle, easily break, and float in water) (Fig. 2-4). They are commonly referred to as *khoai nuoc* ("water taro," or *C. esculenta* var. *aquatilis*). There is variation among these wild taros, and also many ways to use them, together with a remarkable diversity of ingredients. As green vegetables, the edible wild taros are a nutritionally valuable, common food source and are also used as a green fodder for domestic household pigs. Although similar forms are also cultivated, they are much more abundant in the wild than in cultivation. The wild populations might have always been wild, even if people helped spread them.

In Vietnam, the corms of cultivated taros are often cooked with field crabs, pork ribs or other bones, and *Ipomoea aquatica* (a leafy wetland herb related to sweet potato, and widespread in Southeast Asia as a wild and cultivated green vegetable). In northern Vietnam, such soups are popular, especially in the hot summer. To prepare taro stolons, they are peeled, cut into short segments, soaked in salt water, and cooked with fermented rice. This dish is cooling, nutritious, and quite popular. However, after eating, the throat or tongue might still feel a slight itchiness. Culinary folklore tells us that the itchiness will be less if the dish is not touched with bamboo chopsticks while it is cooking (such chopsticks are often used in Vietnamese kitchens). There is similar folklore in the Philippines, warning cooks not to stir taro greens while they are cooking. A chemical explanation for this widespread folk knowledge has not been investigated, but the advice might help reduce the release and dispersion of raphides while leaves and stolons are cooked, or might help maintain high and effective cooking temperatures. Figure 2-8 shows *khoai nuoc* petioles cooked with fish in southern Vietnam.

FIGURE 2-8. Fish stew prepared with young petioles of *khoai nuoc*. (**A**) At left, a wild plant removed from wet ground at the edge of field and forest in limestone hills near the Mekong River; the young stolons have not yet turned green and can grow much longer (at least 1–2 m). At right, young leaves harvested by cutting at top of the corm; some starchy tissue from the corm top remains attached. (**B**) Petioles after washing, peeling off the outer skin, and cutting into short sections. (**C**) Petiole sections placed in bottom of a large pot with water, with pre-cooked eel-fish (from rice-field canals) on top. (**D**) Served with raw green onion (dark green around center), sour lime (lower left), and various condiments in side dishes (not shown). (Photos by P. J. Matthews, southern Vietnam, 2017.)

After boiling on high heat and then simmering, the petioles are completely soft (Fig. 2-8D, light green pieces at center), and have absorbed flavor from the fish.

Bon is a form or cultivar of *khoai nuoc* ("water taro," or var. *aquatilis*) that has little acridity. The Tai people (a minority group in Vietnam) prepare a delicious, cool, and nutritious soup from *bon* petioles that is completely without itchiness (Fig. 2-9). They distinguish *bon* from other wild taros by the presence of a purplish or brown spot at the junction of blade and petiole (Fig. 2-9A; the "piko" in Fig. 2-2). The petioles are collected from wild or cultivated plants, peeled, cut into pieces, and soaked in dilute salted water. Dried buffalo meat or buffalo skin is baked in hot ash, washed, soaked in water until soft, chopped or sliced, and put in a pot to boil until tender. The pieces of *bon* petiole and a leafy green (*Solanum* sp.) are then added, and the cooking continues. When these ingredients are all very tender, further herbs are cut and added: *la lot* (*Piper lolot*) and *rau ngot* (*Sauropus androgynus*). Finally, a mix of ginger and garlic is pounded with some *mac khen* seeds (*Zanthoxylum* sp.) and added. This soup is very suitable in the summer and is a typical Tai dish in northwestern Vietnam.

FIGURE 2-9. Preparation of *bon* soup. (**A**) *Bon* taro in ground. (**B**) Peeled taro petioles, *Solanum* sp. (lower right), and *rau ngot* (*Sauropus androgynus*, lower left). (**C**) *La lot* (*Piper lolot*) and *Solanum* sp. (in basket). (**D**) Chili, *mac khen*, salt, ginger, garlic, and other ingredients. (**E**) Completed soup. (Photos by V. D. Nguyen, northern Vietnam, 2018.)

These examples from Vietnam illustrate the importance of combining local knowledge of wild taro with knowledge of cooking methods and diverse ingredients from wild and cultivated sources. Different varieties of wild taro differ in their acridity despite having similar physical appearance, so what works with the plants in one area might or might not work with plants of similar appearance in another area. In this chapter, we cannot offer specific advice for plants that readers might encounter in their own areas; we offer only general advice on how to appreciate taro as a food and fodder source. We also speculate on possible uses that might encourage more large-scale exploitation in situations where there is a need to limit expansion of wild taro patches.

For any specific area or location, the best source of guidance on how to use the plants is almost always people who live nearby. Wild taros are commensal in villages, towns, and cities across tropical and subtropical Asia, and in other regions. When looking at wild taros, talk to people! Learn from local experience. If local experience is limited, then learn together with the local community, with all necessary caution (see "Acridity" and "Principles for Testing" below).

Acridity: The Gardener's Best Friend

"Acridity" is a word that describes an effect that ranges from a mild itchiness (on the hands when cutting and harvesting plants and peeling taro corms, petioles, or stolons) to an intensely painful "scratchy" sensation on the lips, mouth, tongue, and throat—if raw or poorly cooked taro is eaten, or the wrong plant is used. Acridity is found throughout the Araceae (the plant family of taro, also called "aroids"). Medical reports describe painful results when the sap of ornamental aroids accidentally makes contact with the eye. The recovery is slow (usually days to weeks) but there is no lasting damage. Cows, water buffalos, horses, sheep, and goats are also very sensitive to the effect, judging from their careful avoidance of the plant when grazing the grasses and other herbs that often grow together with wild taro (Fig. 2-5B). In addition to the chemicals that cause acridity, other natural chemicals in taro provide defense against herbivores. Most birds and insects also avoid eating the plant, so taro gardens can be planted with almost no protection. Acridity is definitely the gardener's best friend.

The only large animal that is known to eat taro in the wild and in gardens is wild pig, or boar (Fig. 2-10) (Matthews, 2014; Matthews & Naing, 2005; Matthews et al., 2015; Teron, 2019). This animal is selective, preferring

FIGURE 2-10. Young wild boar feeding at tender base of taro in a garden at night, under a full moon. Woodblock print by Kohno Bairei, Kyoto, 1891. (Photo of print by P. J. Matthews, Japan, author's collection).

young shoots or leaves that are soft and less acrid. The present first author has seen boar damage to the shoots in wild taro patches in Australia, India, and Southeast Asia, and has met farmers in Indonesia and Japan who reported boars entering gardens to feed on taro. Boar "damage" in wild taro patches may stimulate lateral shoot growth and dispersal by the plants, and light harvesting by people may do the same. The young shoots and leaves of wild taros are also what people most commonly harvest and eat. This shared interest in the same wild food plant may have helped lead people to domesticate not just taro, but also the pig. It is remarkable that throughout Southeast Asia, and on many Pacific islands, farmers know that the domestic pig enjoys eating taro, especially when it is cooked and mixed with other ingredients such as rice bran or low-quality, broken rice. Pigs do prefer their taro cooked but appear to become habituated to some level of acridity.

To date, there is no way to directly measure the amount of acridity in taro, before or after cooking. It can only be tested and experienced subjectively, by animal or human. To eat taro requires trust in the cook, or at least a certain amount of experience to be confident about testing taro prepared by ourselves (if we are novice cooks) or by people we do not know. Some principles for testing, and safely gaining experience, are introduced below.

A single idioblast or specialized cell is shown in Figure 2-11. In the leaves and corms of taro, these cells are especially abundant and well developed in older leaf blades, in the epidermis of the leaf stem, and in the skin of the corm. They are scattered singly, and are not strongly attached to other kinds of cells in the tissues where they are embedded. When raw taro tissue is cut or broken, the entire idioblast may be released into a mass of wet broken tissue, or may move out with the sap from cut leaves and corms. After coming into contact with free liquid, the idioblasts swell, burst, and release large numbers of raphides into broken tissue and potentially the mouth of an animal or human consumer. The dense sugar matrix inside the cell quickly absorbs water, causing expansion, a rise in internal pressure, and bursting at each end where the cell walls are thin (Fig. 2-11). The raphides also have small barbs that may help embed them in our skin, tongue, inner mouth walls, and throat, while the long grooves may provide space for the associated proteins (Paull et al., 1999). Similar acridity is found throughout the Araceae.

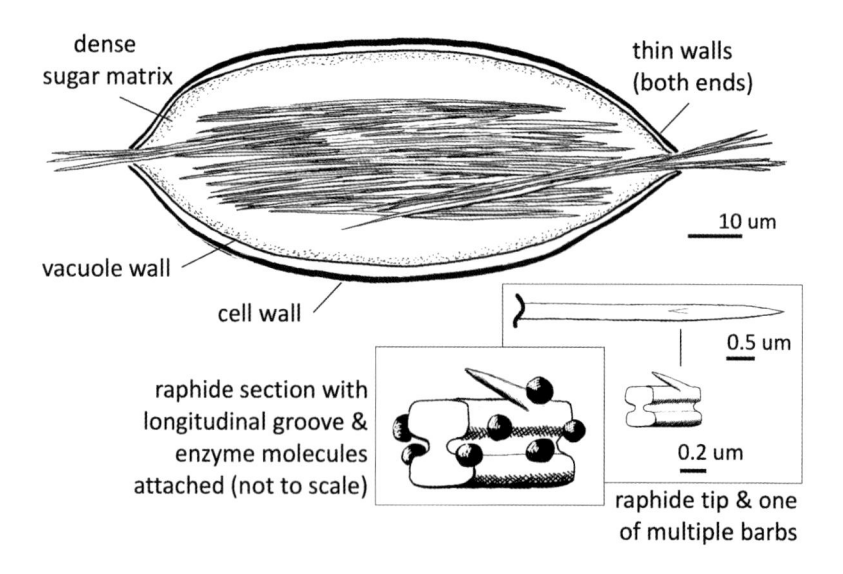

FIGURE 2-11. Idioblasts are the source of acridity (itchiness) in taro. Here a single idioblast cell is shown with raphide bundle, and raphides (needlelike calcium oxalate crystals) ejected at both ends. A vacuole (subcellular compartment with membrane) expands to fill the cell as raphides develop. Acridity is produced by raphides and proteins (enzyme and/or an antigenic small protein) acting together as a complex. The longitudinal groove on each side extends along most of the raphide. Scale bars approximate; cell walls and protein molecules exaggerated in relative size for illustration. (Diagram adapted from Takano & Matthews, 2023; sources: Sakai & Hanson, 1974; Paull et al., 1999, 2022).

GREEN FOOD ZONE: WILD TARO   37

FIGURE 2-12. Below, taro as a naturalized, exotic introduction next to the sacred spring of Argyroupolis on the island of Crete, Greece. Although spreading further downstream, it was appreciated as an ornamental plant, and also remembered as a useful wild food source during war time. Visitors from Cyprus recognized it as the plant cultivated there, and this is supported by vegetative and floral morphology. Above, examples of taro stew (*yiahkni*) served in Cyprus, with all pieces snapped from corms using a knife, and small amounts of lemon juice added during cooking. (Photos by P. J. Matthews, Cyprus, 1996; Greece, 2000.)

All this may sound terrifying, but for thousands of years, farmers, wild plant gatherers, and cooks have harvested and prepared taro for their families and communities, passing on knowledge of how to safely use the plant by example and word. In Cyprus, for example, the ability to prepare a good taro stew (Fig. 2-12) was traditionally considered an important qualification for a potential future wife and family cook (though male cooks can also produce superb taro dishes in modern Cypriot restaurants).

Principles for Cooking

The edibility of taro depends on the (1) variety chosen, (2) parts chosen, (3) timing of harvest, (4) preparation before cooking, (5) other ingredients added during cooking, and (6) heating method (dry vs. wet, temperature, and duration) (Matthews, 2010). Most cookbooks with taro recipes explain just the last three matters (4–6), and this is generally enough when corms or leaves are purchased in a market. The timing of harvest really depends on what kind of food is wanted, and how much.

1. Innumerable (perhaps hundreds) of clonally propagated taro varieties exist, but they have been cataloged only partly, or not at all, in most countries where the crop is grown or naturalized. One variety can be known under many different local names in different languages and different areas, and many varieties can be known under one trade name in one language and area. The names used by farmers are often more specific than the names used by traders and sellers. The latter often indicate nothing more than a general name (e.g., taro) and a country of origin when labeling products for sale. Most wild and naturalized populations have not been described or named at all, though var. *aquatilis* (a stolon-producing variety) was noted in 1960 as widespread in the southern U.S. (Wilson, 1960). In northern New Zealand, it is mainly varieties that produce side-corms that have become wild, and these are known to be good for eating. Recently, in Australia and New Zealand, an ornamental variety of unknown origin, var. *fontanesii*, has naturalized in many locations. This plant has purple-black stolons and petioles, and may or may not be edible: testing is needed (see "Principles for Testing" below).

2. Harvesting an entire leafy top or too many leaves during the growing season will reduce the yield (size and number) of starchy corms. In wild taro populations, the age of plants varies greatly as younger and older generations grow side-by-side in a clump or spreading patch. The younger leaves on younger plants may be most tender and best for eating (Fig. 2-8), if the plants belong to a variety with edible leaves.

3. Different parts should be harvested at different times. If plants in a wild population produce side-corms, then younger side-corms produced within the last year are likely to be best. If a side-corm is sprouting during the growing season, then the young shoot (blade and petiole) and top part of the corm may be good to eat. In mother corms and side-corms, the most starchy and nutritious part is often the upper part and connected base of petiole or shoot (i.e., the entire corm apex area). The youngest part of the shoot is buried deep inside the base of older petioles. When leaves grow, they begin by using starch at the base of the

parent corm; the stored starch is converted into mobile sugars that are carried to the leaves. Each leaf then adds starch to the top of the corm as it reaches maturity. The best time to harvest new corms (mother corms and side-corms) is at the end of the growing season, or afterward. The end of the season is usually marked by whole plants growing more slowly and producing smaller leaves, as temperatures fall or water becomes less abundant. The best time to harvest stolons is early in the growing season, before they become long, hard, and fibrous. The best time to harvest leaves is during the mid- to late-growing season, while growth is still vigorous. If blades are to be eaten, they can be taken when young and fully or partially open, or earlier, while still fully rolled inside the sheath (Fig. 2-2).

4. The main aims of preparation before cooking are to reduce acridity, and to reduce the parts eaten to a size that is comfortable for eating. Corms, side-corms, stolons, and petioles should all be peeled. Many recipes call for corms and side-corms to be peeled before cooking, but peeling after cooking is also a known method that has the advantage of avoiding contact with acrid raw sap. Peeling after cooking is easier, produces a thinner peel, and reduces wastage of the starchy flesh. The cooked peel can be used as animal fodder. If leaf blades are old enough to have thick primary veins (mid- and side-ribs, Fig. 2-2), then the tissue between the ribs should be removed and kept, and the ribs discarded. Preparing blades in this way does not lead to much exposure to liquid sap and can be done with bare hands without much risk of itchiness. Peeling of other parts helps reduce acridity by removing layers where the idioblasts and raphides (Fig. 2-11) are most abundant. These actions expose hands to sap in which raphides have been released. The sap of leaves (petioles and blades) is also especially rich in tannins that go brown on exposure to air, and that can stain clothes permanently. For people with sensitive skin, or who wish to protect sleeves from staining, wearing gloves is helpful (Fig. 2-7D). Staining of clothes is inevitable when removing leaves from corms, or collecting leaves in gardens and wild populations, so it is always advisable to wear old clothes when harvesting taro.

Mature mother corms generally have a thick skin, 1–3 mm thick, which is separated from the starchy storage tissue by a thin layer of cells that give rise to the true roots used by the plant to anchor in the ground and extract nutrients and water from the soil. Roots can grow from anywhere in this middle cell layer, and the growing points, whether active or not, appear as spots when the outer skin is peeled away. This provides a useful marker for complete removal of the outer, highly acrid skin. When most spots are removed, the peeling is deep enough.

There are two very different ways of peeling corms: peeling while washing with large amounts of water, or peeling without any water at all. When water is used, the gummy sap released after peeling a corm is converted into an acrid slime in which raphides are released. Removing the slime requires a large amount of water, and helps remove acridity, but also reduces the nutritional value of corms. Washing with water exposes our hands to acridity, and gloves may be needed if the skin is sensitive. In Egypt, it is common to rub salt into the corm pieces to draw out the sap, and then wash with water. Internationally, using water seems to be most common. In Cyprus, a country that has relatively little rainfall, no large rivers, and little water for domestic use, water is strictly avoided when removing dirt and peeling corms. Dry corms are brushed to remove dirt and then held with a towel while peeling and cutting with a knife. This minimizes the release of sap and the exposure to acridity during peeling and cutting and retains all the nutritional value of the corm. In both schools, corms are cut down to mouth-sized pieces by either slicing the corm with a knife, or (in Cyprus, Fig. 2-12) snapping pieces off by inserting a knife blade and using leverage. Since wet peeled corms are very slippery, it is important to take special care when cutting them (a dry towel helps provide grip).

When peeling petioles and stolons, the thin outer skin is removed with fingers alone or with the help of a knife edge. Peeling also helps reduce acridity, which is more concentrated in the skin, as in corms. Fresh stolons are brittle and can be peeled and snapped by hand to produce short pieces (Figs. 2-3A, 2-7D). After peeling first, the petiole can be cut into short pieces (Fig. 2-8B). Washing the peeled petioles or stolon pieces in water may also help reduce acridity.

Another important method for reducing acridity is drying the edible parts after they have been peeled or sliced. Corms can be cut lengthways and partially dried, and peeled petioles can be split lengthways to near one end, then placed on a string in shade to dry slowly with airflow. Intact young leaves, with tender petioles and blades, can also be air dried in bundles, or leaf tissue stripped away from the ribs can be dried, resulting in many small light pieces that are suitable for soups (a common practice in the Philippines).

Light fermentation of raw, sliced corms and peeled petioles, with salt and other ingredients, may also help reduce acidity before cooking (Q. Fang, pers. comm., and Matthews, field notes, China, 2018). *Poi*, a fermented starch paste made in Hawai'i and other Pacific islands after pounding cooked taro, is not at all acrid. Drying and fermentation (as silage) have both been used experimentally to prepare animal fodder from raw taro leaves and corms, with the goal of extending the storage life of taro fodder, for its consumption by pig (Kaensombath, 2012) and animals that do not usually eat it (e.g., chickens, goat, sheep, water buffalo). Fermentation in taro silage removed acidity by acting on components other than the raphide crystals (Carpenter & Steinke, 1983) (see also "Future Prospects" below).

5. Other ingredients used when cooking taro have two main functions: improving taste, texture, and appearance, according to personal or common cultural preferences, and adding nutritional value. This is a huge and largely unexplored subject in taro research. From the experience of eating taro corms in many different countries, it is clear that taro has an excellent ability to absorb flavors from many different ingredients and combine them with the mild taro flavor and soft texture in many pleasing ways. An additional function of other ingredients can be to reduce acridity, by chemical effects on the raphide-protein complex shown in Figure 2-11. Boiling in water with a raised pH (alkaline) helps dissolve the calcium oxalate crystals, and the pH can be raised by adding baking soda (sodium bicarbonate) (Matthews, 2010; Kumoro et al., 2014). Adding lemon or lime juice lowers the pH, making a more acidic solution—a common practice in Cyprus and Vietnam to adjust the flavor and texture of cooked taro (Figs. 2-8D, 2-12). This helps to reduce or disperse the slime released when corm sap contacts water used for cooking. The chemical effects of cooking methods on taro have not been well studied. Other ingredients may also protect the skin in our mouth and throat by creating a protective (and flavorful) layer of protein and fat. The prime candidate for this form of acridity-reduction is coconut milk, an ingredient that combines extremely well, in flavor terms, with taro corms and greens in many dishes that are popular in Southeast Asia, South Asia, and Oceania.

6. Heating methods used when cooking taro include baking, frying, boiling, steaming, and heating with higher or lower temperatures, with or without high pressure (in a pressure cooker), for shorter or longer periods. Baking, frying, and heating with water (boiling, simmering, steaming) all produce different taste and texture effects. Cooking on hot stones in a pit (*umu, hangi,* and other names) is common in Oceania, and combines the effects of both baking and steaming, under a cover of leaves and earth; the method is very efficient (in energy and labor terms) for large quantities of food. For starchy corm pieces and green parts (blades, petioles, stolons), a key marker that cooking has been sufficient to remove acridity and absorb flavors is softness. Softness can be tested with any sharp or pointed utensil. Old corms that remain hard may be not cooked enough to remove acridity, or may lose acridity but resist becoming soft and flavorful because they are fibrous and lack starch. Recipes published by authors familiar with varieties that produce small corms (eddo types) may recommend cooking times that are too short for larger corms or cut pieces. A Japanese recipe for a variety with small side-corms or cut pieces (less than 50 g each, for example) may need boiling for only 10–15 minutes, which is far too short for larger pieces, or entire corms weighing up to 1 kg or more. The care and time needed to prepare wild taro in Queensland, Australia (R. Tucker in Matthews, 2014), is much greater than for any cultivated taros, or for the commensal wild taros that are commonly eaten.

Given the wide range of variables outlined above, how is it ever possible to know if a particular taro variety and dish can be eaten? Even in families who regularly cook taro, it is natural for the cook to be responsible for checking that suitable raw cooking material is used and that the resulting dish is ready to eat. When familiar taro varieties are cooked with familiar methods, minimal testing is needed and is no more difficult than for other foods. In all areas of cooking, domestic and professional, testing by taste is common, but acridity in unfamiliar taro varieties poses a special problem.

Principles for Testing

The terms "acridity" or "itchiness" refer to a sensation that can only be perceived subjectively, using our own sensory abilities. When people are asked to taste taro in a cooking trial, the group asked to do this work is called a "sensory panel." When researchers offer taro to animals as food, it is called a "bioassay." For obvious practical and ethical reasons, care must be given to ensure that members of a sensory panel do not have dangerous experiences (not all research reports demonstrate such care, but most do). The same care is needed in non-academic, informal settings, when exploring the possible uses of wild taros with family members, friends, and local communities. "Informed consent" and "volunteer" are keywords for the ethical approach, as well as "safety."

Traditional (unwritten) methods for teaching children or cooks about acidity have not been reported, and many cookbooks with taro recipes do not provide advice on the matter. When cooking an unfamiliar taro for the first time, please cook the corm or pieces thoroughly, until all parts are soft. Five levels of testing are introduced next.

1. Inner wrist (lower forearm) test. The safest subjective test that can be carried out uses soft skin of the inner wrist or forearm as a target area (Paull et al., 1999). This skin is too insensitive to distinguish completely and partially cooked taro (both with reduced acridity), but is not too sensitive for testing completely uncooked (raw) taro tissues. The possibility of strong allergic reactions cannot be excluded because large clinical trials have never been conducted with taro. Such reactions must be rare, since millions of cooks have cleaned and prepared taro without plastic gloves and without ill effect for thousands of years, despite surely experiencing acridity on hands and wrists. For most cooks, the reactions are mild and tolerated, or are avoided by careful handling in the kitchen, or by handling only familiar low-acridity forms of taro. People with strong skin reactions, or little tolerance, may choose not to cook or eat taro, and informants have reported making such choices in countries where taro is commonly eaten (Matthews, field notes).

    The inner wrist test has been used personally many times by the present first author, and also with occasional volunteers: Sap from a small quantity of cut, mashed, or crushed tissue (corm, petiole, or blade tissue) is rubbed onto an area of ca. 4 cm^2 on soft skin of the inner wrist. After a short period (1–2 min.), an itchy sensation may or may not be noticed. The waiting period and the intensity of itching vary according to source material and the sensitivity of each individual. Itching usually becomes gradually less, and disappears within several minutes, or less than 30 minutes. An antihistamine (insect-bite) cream can be applied to treat and remove the itchiness if wanted (Matthews, field notes). Hands and fingers that have been in direct contact with sap should be washed afterward to avoid rubbing the eyes with traces of raw sap.

The same wrist test after cooking taro can be used to demonstrate an overall reduction in acridity, but does not have enough sensitivity to confirm that the taro is good to eat. If the test shows that obvious acridity has been removed, a taste test can be considered. Volunteers who have not previously eaten taro but who experience wrist tests with raw and cooked taro might more easily accept that cooking is effective, and can be more confident that the effect will wear off if acridity is revealed by tasting while cooking the plant.

2. Tip of tongue test. The safest test for tasting is carried out on the tip of the tongue, with just enough material to detect the physical presence. Briefly hold a piece of cooked taro between the tongue and front teeth, just long enough to gather saliva and spit out the material. If there is no unpleasant effect within five minutes, it is safe to proceed to the next test.

3. Tongue body test. This will reveal more flavor from the cooked taro and any other ingredients that have been included. Hold a small amount of cooked taro briefly between the middle or front of tongue and mouth wall, then discard. If there is no unpleasant effect within 5–10 minutes, then proceed.

4. Gentle chewing test. Again, hold a small amount of cooked taro on the tongue, near the front (away from throat), then chew gently and briefly before spitting. If there is still no sign of acridity within 5–10 minutes, then proceed.

5. Swallowing test. Gently chew and swallow a small quantity, then wait a few minutes to check for any itchiness in the throat. Mild itchiness in the mouth and throat will soon disappear, but drinking milk can help reduce the effect if it is too unpleasant.

The Level 5 test may be the only test needed by cooks when using a familiar wild or cultivated taro to prepare a familiar dish. It can also be used to test for texture and flavor when there is no expectation of itchiness.

In summary, the following series of tests are recommended for cooked taro, all with very small amounts, and starting with Level 1.

Level 1. Inner wrist test (also safe with raw or undercooked taro).
Level 2. Tip of tongue, touch and spit.
Level 3. Tongue body, touch and spit.
Level 4. Gentle chewing, then spit.
Level 5. Gentle chewing, then swallow.

These are all tests that are ideally carried out together with someone who is already experienced with identifying, cooking, testing, and eating taro. The present first author has experienced acridity in the mouth that lasted for several hours, despite thorough cooking, after Level 4 testing of a wild form of *Leucocasia gigantea* (formerly known as *Colocasia gigantea*). Never eat an unknown plant that simply "looks like taro." Many wild and ornamental aroids (*Alocasia* spp., *Caladium* spp., *Colocasia* spp., *Leucocasia gigantea*/formerly *C. gigantea*, *Xanthosoma* spp., and others) resemble taro (*C. esculenta*) in part, but may have greater acridity. The authors of websites and other social media often misidentify taro in photographs, confusing it with other species and genera. This can happen in specialist botanical or cooking websites, but is most common in personal or informal commercial media concerned with gardens or food. Please choose information sources carefully, and check more than one source, when using photographs or diagrams to identify unfamiliar plants that might be taro, or when publishing a recipe.

Future Prospects

As taro becomes more familiar to more people, and its beneficial qualities are more widely appreciated, efforts to eradicate wild populations will be able to focus on situations where (a) natural ecosystems are being protected or restored, and/or (b) wild plants belong to a variety that is unsuitable for eating and are not part of the natural flora. Since taros disperse vegetatively from upstream to downstream, removal of plants should be most effective if it begins at the highest point of occurrence in a catchment and proceeds downstream. In some areas it may be possible to use satellite image analysis (Everitt et al., 2007) to locate the largest upstream populations. This will reduce the possibility or speed of reinvasion from upstream to downstream. The same principle can be applied by landowners who wish to keep their plants contained in a defined area. Unintended movements downstream from taro cultivations can be limited by careful attention to the disposal of discarded or unwanted parts. For wild populations that are not already utilized, relatively simple DNA tests can be applied to distinguish likely edible Clade I or Clade II lineages from the likely inedible Clade III wild taros identified by Ahmed et al. (2020).

Developing non-sensory proxy tests for acridity, by measuring chemical reactions of the acridity complex, would make it possible to grade different wild and cultivated taros and apply suitable cooking methods, and would help to reduce anxiety when faced with unfamiliar plants. Naturalized wild populations that are derived from cultivated forms after past introduction are well-suited for exploration as food sources, especially in regions where natural breeding (seed production and germination) does not occur. In the absence of breeding and genetic reassortment, wild taros keep most of the food qualities for which they were originally introduced, though food qualities do vary according to the growing conditions outside cultivation.

Non-food uses for wild taros may also gain greater favor in the future. In Vietnam, when the country was still very poor, the Vietnamese government strongly promoted the use of taro as a fodder for domestic pig—building on what was already an ancient tradition in that country and tropical Asia generally. This may have contributed to the present abundance of edible wild taro populations (and domestic pigs) in many parts of Vietnam. To manage and make better use of wild taro populations generally, we need to learn more about their genetic diversity, modes of reproduction, dispersal history, cultural uses as food and fodder, and ecological interactions in diverse environments. Removing invasive populations outside the natural range of taro may help restore habitats needed by useful native plants, or might create opportunities for other less useful invasive plants to become dominant. To manage wild and cultivated plant communities responsibly, we must balance many different goals. As a preemptive measure, future development and dissemination of taro as a food crop or ornamental plant could give priority to cultivars that have limited dispersal ability (e.g., cultivars that do not produce fruit and seeds, or long and easily broken stolons). Such cultivars are already dominant in temperate regions (Matthews, 2014; Ahmed et al., 2020). They are sterile triploids that produce only side-corms, similar to the cultivar shown in Figure 2-6. In temperate regions, taro is rarely invasive over long distances or in large, rapidly spreading patches.

In conservation areas where there is a clear need or wish to remove or reduce exotic taro populations, it might help if economically beneficial uses can be found to help cover the costs of removal. Fodder uses may have the greatest potential in this regard, and might provide some revenue with which to fund management efforts. Taro may also be good as fodder for wild-caught boar or their young, as a specialty meat source, in areas where boar are invasive. Taro is a natural food for pig, and the abundance of tannins in the leaves may also make it a medicinal food for this animal. Farmers in Southeast Asia find that taro helps keep their household pigs healthy. Traditional diets that include taro do not produce fast-growing, large animals, but do produce animals with high-value meat. The role of taro as a fodder in Southeast Asia may be similar to the role of tannin-rich acorns in the diet of pigs in western Europe (resulting in high value "Iberian pork," for example). Bringing taro into modern animal production systems might be beneficial for the animals and also for people, especially if it can help reduce dependence on the use of antibiotics in animal husbandry. In any case, there is definitely potential to use wild taros to develop new fodder products (dried leaf pellets, fermented silage) that can be stored and used on a larger scale. Other invasive plants also have value as fodder (e.g., kudzu vines, paper mulberry trees, American oaks in Europe, and some grasses). Using them in this way might improve animal health, reduce the use of arable land for grazing and fodder production, and provide economic incentives for managing invasive plants as early colonizers (successional vegetation) in erosion control, forest regeneration, and agroforestry.

To conclude: welcome to the green food zone! The present authors have been guided by many others (Matthews et al., 2017). When approaching wild taros for the first time, talk to people, take care, and enjoy the journey. There is much to learn.

Acknowledgments. The first author thanks the many friends, colleagues, institutions, and community members who have encouraged and supported his efforts to learn about wild taros, including D. Hunter (Australia), M. Hasanuzzaman (Bangladesh), C.-L. Long, Q. Fang (China), P. Croft (Cyprus and Greece), A. Dasgupta, D. K. Medhi (India), T. Akimichi, K. Ikeya, T. Masuno, E. Tabuchi, E. Takei, Y. Sato, K. Watanabe (Japan), K. W. Naing (Myanmar), N. Araho, J. Waki (Papua New Guinea), E. Maribel Agoo, D. Madulid, D. Tandang, M. Medicilo (Philippines), D. Sookchaloem (Thailand), and V. K. Nguyen (Vietnam). Grants supporting the present work were: JSPS Kakenhi Nos 17H04614 and 17H01682 (Matthews), Vietnam National Foundation for Science and Technology Development (NAFOSTED) 106.03-2019.322 (Nguyen).

Literature Cited

Ahmed, I., P. J. Lockhart, E. M. G. Agoo, K. W. Naing, D. V. Nguyen, D. K. Medhi & P. J. Matthews. 2020. "Evolutionary origins of taro (*Colocasia esculenta*) in Southeast Asia." *Ecology and Evolution* 10: 13530–13543. https://doi.org/10.1002/ece3.6958

Candolle, A. de. 1885. *Origin of Cultivated Plants*. Kegan Paul, Trench and Co., London.

Carney, J. A. & R. N. Rosomoff. 2009. *In the Shadow of Slavery: Africa's Botanical Legacy in the Atlantic World*. University of California Press, Berkeley and Los Angeles.

Carpenter, J. R. & W. E. Steinke. 1983. "Animal feed." Pp. 269–300 *in* J.-K. Wang (editor), *Taro: A Review of Colocasia esculenta and Its Potentials*. University of Hawai'i Press, Honolulu.

Catesby, M. 1781. *Natural History of Carolina, Florida, and the Bahama Islands*. London.

Everitt, J. H., C. Yang & M. R. Davis. 2007. "Mapping wild taro with color-infrared aerial photography and image processing." *Journal of Aquatic Plant Management* 45: 106–110.

Grimaldi, I. M., S. Muthukumaran, G. Tozzi, A. Nastasi, N. Boivin, P. J. Matthews & T. v. Andel. 2018. "Literary evidence for taro in the ancient Mediterranean: A chronology of names and uses in a multilingual world." *PLoS One* 13: e0198333.

IPGRI. 1999. *Descriptors for Taro (Colocasia esculenta)*. International Plant Genetic Resources Institute, Rome.

Kaensombath, L. 2012. *Taro Leaf and Stylo Forage as Protein Sources for Pigs in Laos: Biomass Yield, Ensiling and Nutritive Value*. Swedish University of Agricultural Sciences, Uppsala.

Kumoro, A. C., R. D. A. Putri, C. S. Budiyati, D. S. Retnowatia & Ratnawati. 2014. "Kinetics of calcium oxalate reduction in taro (*Colocasia esculenta*) corm chips during treatments using baking soda solution." *Procedia Chemistry* 9: 102–112.

Matthews, P. J. 2006. "Written records of taro in the Eastern Mediterranean." Pp. 419–426 *in* Z. F. Ertug (editor), *Ethnobotany: At the Junction of the Continents and the Disciplines (Proceedings of the IVth International Congress of Ethnobotany, ICEB, 21-26 August, Istanbul, Turkey)*. Yayinlari, Istanbul.

Matthews, P. J. 2010. "An introduction to the history of taro as a food." Pp. 6–30 *in* V. R. Rao, P. J. Matthews, P. Eyzaguirre & D. Hunter (editors), *The Global Diversity of Taro: Ethnobotany and Conservation*. Biodiversity International, Rome.

Matthews, P. J. 2014. *On the Trail of Taro: An Exploration of Natural and Cultural History*. National Museum of Ethnology, Osaka.

Matthews, P. J. & K. W. Naing. 2005. "Notes on the provenance and providence of wildtype taros (*Colocasia esculenta*) in Myanmar." *Bulletin of the National Museum of Ethnology* 29: 587–615.

Matthews, P. J. & M. E. Ghanem. 2021. "Perception gaps that may explain the status of taro (*Colocasia esculenta*) as an 'orphan crop.'" *Plants People Planet* 3: 99–112. https://doi.org/10.1002/ppp3.10155

Matthews, P. J., V. D. Nguyen, D. Tandang, E. M. Agoo & D. A. Madulid. 2015. "Taxonomy and ethnobotany of *Colocasia esculenta* and *C. formosana* (Araceae): Implications for evolution, natural range, and domestication of taro." *Aroideana Supplement* 38E(1): 153–176.

Matthews, P. J., P. J. Lockhart & I. Ahmed. 2017. "Phylogeography, ethnobotany, and linguistics: Issues arising from research on the natural and cultural history of taro *Colocasia esculenta* (L) Schott." *Man in India* 97: 353–380.

Matthews, P. J., V. D. Nguyen, Q. Fang & C.-L. Long. 2022. "*Colocasia spongifolia sp. nov.* (Araceae) in southern China and central Vietnam." *Phytotaxa* 541: 1–9.

Paull, R. E., C.-S. Tang, K. Gross & G. Uruu. 1999. "The nature of the taro acridity factor." *Postharvest Biology and Technology* 16: 71–78.

Paull, R. E., D. Zerpa-Catanho, N. J. Chen, G. Uruu, C. M. J. Wai & M. Kantar. 2022. "Taro raphide-associated proteins: Allergens and crystal growth." *Plant Direct* 6: e443.

Safford, W. E. 1905. *The Useful Plants of the Island of Guam*. Government Printing Office, Washington.

Sakai, W. S. & M. Hanson. 1974. "Mature raphid and raphid idioblast structure in plants of the edible aroid genera *Colocasia*, *Alocasia*, and *Xanthosoma*." *Annals of Botany* 38: 739–748.

Takano, K. T. & P. J. Matthews. 2023. "Mystery of taro." *Chiiki Bunka* [*Local Culture*] No. 144 (Spring): 22–23.

Teron, R. 2019. "Ethnobotanical study of dietary use and culinary knowledge of aroids (Family Araceae) in Karbi Anglong district, Assam." *NeBIO* 10: 80–84.

Vehling, J. D. 1977. *Apicius: Cookery and Dining in Imperial Rome*. Dover, New York.

Wilson, K. A. 1960. "The genera of the Arales in the southeastern United States." *Journal of the Arnold Arboretum* 41: 47–72.

CHAPTER 3

A Brief Biography of Itadori: Re-storyation as an Approach to Japanese Knotweed Management

Tusha Yakovleva

1. Home Life

The Japanese Archipelago, eastern coast of China, Taiwan, and Korean Peninsula are prone to unpredictable disturbance. The humid continental climate, combined with a strong maritime influence, is a hospitable environment for a great diversity of plants along with the rich diversity of human cultures who depend on them. The generous soils and seasons, however, can give way to volcanic eruptions, tsunamis, and earthquakes, evacuating all beings who make home there out of their familiar existence on extremely short notice. For all inhabitants of these mercurial places, acknowledgment of and adaptation to sudden change in circumstance are essential to livelihood. For many plants, just like for people, living a full and resilient life in climatically unpredictable habitats means learning to hold two opposing forces in symbiosis. On one hand, setting strong roots and taking refuge in community are important for living through sudden and dramatic change. On the other hand, uprooting at a moment's notice is also essential to survival.

This seismically active region, much of which sits within the Pacific Ring of Fire, is home to two closely related species of Itadori (in Japanese) or *Reynoutria japonica* Houtt. and *R. sachalinensis* (F. Schmidt) Nakai of the family Polygonaceae (in binomial nomenclature) who have mastered the art of living with uncertainty: staying strongly rooted while simultaneously capable of relocating in the blink of an eye. Within their[1] home range, Itadori habitats are somewhat split, further pointing to their impressive adaptability. Itadori set root both in sunny areas at higher elevations, including mountainous lava field scree surrounding active volcanoes, as well as in fertile, moist bottomlands such as river floodplains, where they live alongside similar fast-growing, robust plants like bamboo (*Sasa* spp.) (Child & Wade, 2000).

The genus *Reynoutria* comprises seven or so species, all of whom are herbaceous, generally monoecious, rhizomatous perennials. Other scientific names that have been widely used for *R. japonica* are *Fallopia japonica* (Houtt.) Ronse Decr. and *Polygonum cuspidatum* Siebold & Zucc. *Reynoutria japonica* and *R. sachalinensis* rely primarily on vegetative reproduction, forming dense, clonal colonies that remain rooted for many years (Shaw, 2013). However, to adapt to the ever-present possibility of dramatic disturbance, Itadori prepare for this with a number of adaptations. Their powerful, rhizomatous root system is capable of supporting the growth of new shoots for multiple dry seasons, with woody rhizomes that can grow 10 feet down into the ground and exceed 40 feet in lateral growth (Pridham et al., 1966). Being herbaceous perennials means Itadori don't have to worry much about protecting sensitive aerial buds during a sudden disturbance event or drop in temperature and, once conditions for growth are right again, can grow up to eight inches per day (Barney et al., 2006). Not only are Itadori quick in growth, the shoots are strong enough to push through hard surfaces, such as highly compacted flooded soils or lava scree. Plus, Itadori's tenacious vegetative reproductive abilities allow the plants to regrow from tiny root nodules, which is particularly advantageous in the case of a flood, when bits of unearthed Itadori rhizome fragments, as small as one-half inch, can travel to new suitable locales and regrow. The rhizomes can survive below the soil surface in a dormant state for up to 20 years, while the plant's allelopathic

---

1. As Itadori is an animate being, having more in common with you and me rather than a dusty shoe, it is only right to address the plant in a way that acknowledges their aliveness. Since the English language does not naturally allow this, I have chosen to refer to Itadori or Knotweed—a plant who blurs the boundaries between an individual and the collective—by the plural pronoun "they."

properties assist young Itadori in establishing themselves in a new growth site (Conolly, 1977; Locandro, 1978).

Of course, to root in community, one must not only meet their own needs for survival but also coexist with neighbors. Itadori are entangled in diverse community relations. Among the relations the plants form with others with whom they share a home are over 180 native insect species (Shaw et al., 2011) who feed on Itadori and help keep the plant population in balance. They co-conspire with an obligate psyllid or jumping plant louse, *Aphalara itadori*, who feeds only on Itadori. Itadori's tangled roots weave soils back together after volcanoes erupt as well as aerate soil after flooding and other compaction events. Itadori's fast-growing, abundant foliage not only adds organic matter to thin soils, but also has a particular knack for uplifting subterranean minerals, making them bio-available to other plants after senescence. In their early-succession pathfinding ways, Itadori ameliorate the soil enough to eventually make way for other plants (Yoshioka, 1974). Meanwhile, in the floodplains, Itadori often share space with other perennials, primarily other tall forbs.

Itadori have also supported their human neighbors since time immemorial. They contributed to the diet of Okhotsk culture (Leipe et al., 2017) and continue to offer many gifts to the Ainu, Indigenous peoples of Hokkaido, Sakhalin, and Kuril islands (Crawford & Yoshizaki, 1987). *Reynoutria japonica* is known as Kuttar in Ainu, while *R. sachalinensis* is called Ikokuta, Ikokuy, or Irurex. Kuttar's tender young shoots provide one of the first fresh greens of the season and are eaten raw or in stews, traditionally along with crushed Kamchatka lily (*Fritillaria camschatcensis*) bulbs, while the seeds are ground up and added to grains and stews (Chiri, 1953; Ohwi, 1984). Kuttar also offer an ethnobotanical resource to support another important food in Ainu diets: starch-rich roots (starch is not readily abundant outside cultivation). "Kuttar" means "tube" in a variant of Ainu and their mature woody stems provide a tubular storage container and cooking vessel for ground-up starchy bulbs of varied plants (Kayano, 1996). In Ainu culture, Kuttar stems are used both to transport this important staple food and to bake it (inside the Kuttar stem) in a fire (Williams, 2017).

The dry stems also serve as wind and snow breaks along austere parts of island coasts, while large leaves offer wrapping for smoking herbs (Chiri, 1953). Kuttar are also categorized among the moskar, meaning "to cut grass," various sturdy plants that are used as material for impromptu mats to hold fish or meat (Chiri, 1953). On Sakhalin, Ikokuta are also the star of many winter meals, which are built around cikaripe, a seasoned root or stem, often

Reynoutria sachalinensis, to which other ingredients including dried or frozen berries, seaweed, and salmon roe are added (Williams, 2017). Ainu peoples also turn to Ikokuta for medicine: the plant's disinfectant leaves prevent boils (Mitsuhashi, 1976); roots are applied topically to treat ringworm and scabies, and taken internally to treat sexually transmitted diseases and diarrhea (Chiri, 1953; Butler, 1994).

In Japanese cultures, Itadori—which has been translated as "take the pain away"—are honored as both food and medicine. As medicine, Itadori leaves are used as a topical rub to take away pain from open wounds and sores (Williams, 2017). In the culinary realm, Itadori's place is among the sansai, wild mountain vegetables (largely shoots and leaves) that are traditionally collected in spring, often gathered in *satoyama* landscapes, to celebrate the arrival of a new season and offer missing nutrients to winter-weary bodies (Indrawan et al., 2014). The shoots are gathered while tender, before they turn woody, and, as with other shoot vegetables, have a brief harvest window: a mere two to three weeks. Itadori are served fresh in spring but occupy a more significant place in Japanese kitchens as preserved vegetables. Tsuruoka-based forager Alexis Crump explains that "Itadori is traditionally picked in spring and summer, then eaten during the winter months. This is because it is necessary to heavily salt and cover it for several months before it reaches a consistency that is edible. Thanks to this process, it also acted as a source of food during a time when there wasn't much food available for people who lived before electricity. . . . It also shows up in local school lunches during the winter as well" (Crump, pers. comm., 2022). Itadori are also within the constellation of sacred foods of the Shojin Ryori, a place and plant-based monastic cuisine that is part of purifying Yamabushi practices (Kobayashi, 2004).

In China, Itadori—known as Hu Zhang—have a revered role in Traditional Chinese Medicine in treating cancers (Kimura & Okuda, 2001); as powerful allies for fortifying the body's immune response to viruses, microbes, and inflammation; and as protectors of the nervous system and heart (Peng et al., 2013; Zhang et al., 2013). Hu Zhang also have a long history as a source of fabric and food dye (Liu et al., 2014). The cultural relations between Hu Zhang and people are as diverse as the cultures with whom the plants share homeland: the ones described here are but a sliver of Hu Zhang stories passed from one generation of humans to the next.

Since time immemorial, Itadori, their people and their many interspecies kin coexisted in reciprocity with one another; Itadori offering gifts of nutrition, medicine, material; while psyllids, fungal diseases, volcanoes, and other evolutionary co-conspirators offer the gift of keeping Itadori landscapes in balance. In parts of Itadori's home range, such intertwined existences are continuing, following biocultural pathways known to those lands (Shimoda & Yamasaki, 2016).

2. Novel Dispersal

In the 1820s, a novel dispersal mechanism, perhaps less predictable than seismic disturbance, swept into Itadori's homelands: imperial botany. In *Land of Plants in Motion*, Thomas Havens describes this particular longing for Chinese, Korean, and Japanese flora as "a quest to bring the world's vegetation home for the delectation of the overseers of empire" (Havens, 2020). European plant explorers were tasked with bringing living plants from around the world back home to feed growing curiosities for the aesthetic as well as the scientific mystique of foreign vegetation, while simultaneously feeding growing commercial possibilities of international plant trade.

The English bourgeois developed a taste for growing "outdoor pleasure grounds," private parks planted out with species from around the planet. As Havens (2020) explains:

> A hunger for unseen plants, animals, and artifacts was piqued by accounts from far-flung travelers and reports by Jesuit missionaries abroad, triggering the emergence of collecting as a pastime. The eighteenth-century enlightenment fed an appetite for encyclopedic knowledge and led to a new cosmopolitanism of curiosity about the whole world. In a newly secular age, the rise of scientific inquiry expanded the scope of investigation to the entire planet, revealing that plants knew no borders and were in constant motion.

Philipp Franz von Siebold was a German physician with a passion for travel and natural sciences, whose travels to Japan corresponded with the blossoming enthusiasm across Western Europe for plants from East Asia. Siebold moved plants and other beings out of their home ecosystems with fervor, introducing around 1,000 new plant species from Japan across Western Europe via his mail-order nursery trade (Havens, 2020). The frenzy of the European elite for gardening with plants from foreign ecosystems quickly reached those with wealth and want of social status in the so-called United States. Horticulture became a popular pastime and within a few decades, the descendants of Siebold's plant transplants reached the soils of Turtle Island, the continent otherwise known as North America.

Itadori were among the plants Siebold plucked out of their home ecosystems and promoted across Western Europe. In Europe, Itadori took on yet another name: Japanese Knotweed. By 1850, Knotweed were being planted in European "pleasure grounds," recommended by famous ornamental gardeners of the day for their large stature, hardiness, and quick, vigorous growth (Robinson, 1870; Wood, 1884; Earle, 1897). Soon, Knotweed were also being welcomed by cattle farmers, who grew the plant as a nutritious boost for animals in the shoulder season before grasses were ready for grazing. By the start of the 20th century, Knotweed were planted in the newly formed New York Botanical Garden, located in Lenape homelands, otherwise known as New York City (Del Tredici, 2017).

Japanese Knotweed were welcomed with open arms in Europe and on Turtle Island not just by the handful of humans who decided to promote the plant, but even more so by the relatively mild climates, abundant fertile soils, ample moisture, an absence of 180 ancestral companion species with a taste for Knotweed to keep apace with, an absence of equally robust herbaceous perennials to share space with, no familiar plant and fungal diseases to wrestle with, far fewer large-scale dramatic natural disturbances such as volcanic ash and landslides to slow their growth, and—crucially—endless upturned soils (a by-product of development) to germinate in. In these new conditions with few natural checks, Knotweed grew beautifully. The plant thrived and spread; by early 20th century, Knotweed naturalized exponentially in Europe and was on a similar trajectory in northeastern Turtle Island. The rate and success of Knotweed's establishment in unfamiliar soil so impressed plant ecologists that it led some to conclude that "Japanese Knotweed is actually much more 'at home' abroad" (Shaw, 2013).

Since Knotweed rely on vegetative reproduction and with great success, the plant's dispersal across Turtle Island and Western Europe was conducted almost exclusively by people (Dauer & Jongejans, 2013). In 2000, an analysis of Japanese Knotweed (*Reynoutria japonica*) populations around the world outside of their home range revealed that the majority of Japanese Knotweed growing in Europe and Turtle Island are clones of the single plant brought to the Netherlands by Siebold in 1849 (Hollingsworth & Bailey, 2000).

For many years, Knotweed continued to be the darling of imperial botany aesthetic, sold by many nurseries for ornamental landscaping. Additionally, Knotweed were, for a time, called upon by restoration ecologists to aid in erosion control and planted intentionally along eroding riverbanks. Knotweed's expansion was also greatly powered by inadvertent transplanting to new locations through the movement of soils containing Knotweed root fragments from countless construction projects like building foundations, ditches, levees, highways. In recent decades, Knotweed's flourishing has also been fueled by climate change–driven flooding and warmer winters as well as accompanying water pollution. Through shoreline disturbance from flooding, Knotweed has dispersed along many miles of fresh and brackish shorelines; increased salinity and eutrophication, which Knotweed can tolerate while many other plants cannot, worked in consort to help more Knotweed establish (Shaw, 2013). Roadways, with their endless linear ditch wetlands, frequent soil disturbance, and high salinity, made for yet more perfect Knotweed habitat.

Knotweed's presence became hard to ignore. The plant's adopted range spread in step with a burgeoning new branch of environmental science: Invasion Biology, which offered ecologists new militant language for describing the movements of organisms outside their known homeland. Scuffles between Knotweed and people flared up, both in the United States as well as Europe, with the very qualities that not long ago attracted horticulturalists, farmers, and ecologists to Knotweed—quick, vigorous growth and impressive indifference to compact, polluted, saline, or infertile soils—now raising alarms. Among Knotweed's primary offenses was and continues to be their ability to grow through concrete, a talent Knotweed perfected through coevolution with lava fields (Fennell et al., 2018). In addition, Knotweed were now being blamed for causing stream bank erosion—by crowding out other plants and thus ostensibly leaving soils bare in the fallow season—the very thing the plant was once thought to fix (Gilbert, 1992).

Built environments—public infrastructure and private homes alike—were now permeable to Knotweed (Fennell et al., 2018). Fear quickly bloomed into blame, and Knotweed were deemed a *terrorist*, charged with attacking homes and communities. In the United Kingdom, stories began circulating of mortgages being denied on lands with Knotweed present, thus driving down real-estate values (Shaw, 2013). Lawsuits sprung up between neighbors over the presence of the plant, in a desperate effort to somehow quantify Knotweed's offenses. An industry rose up to defend against Knotweed's influences, utilizing any measures to eradicate the plant. Glyphosate—a powerful herbicide that has been associated with increased risk of lymphoma and autism spectrum disorder (Andreotti et al., 2018; von Ehrenstein et al., 2019) and has harmful effects on honeybees (Tan et al., 2022)—became the Knotweed eradicator's primary partner. However, Knotweed's roots are powerful too and proved stronger than glyphosate in the majority of cases (Kabat et al., 2006). A blunt physical approach to eradication, an alternative to chemical measures, was and continues to be used too: scraping soil clean with heavy machinery in an attempt to clear out Knotweed root mass in its entirety. This greatly disturbs a habitat, including potentially altering the hydrology of the area; moreover, the efficacy of this approach is unreliable, as it is necessary to precisely dig out every fragment of Knotweed's knotty underground biomass (Baker, 1988; Seiger & Merchant, 1997). Without repeated applications, Knotweed are likely to reestablish anew. Other experimental eradication tactics in consideration are biological controls, such as the introduction of a known fungal disease from Knotweed's first homelands or the psyllid (*Aphalara itadori*) for whom Knotweed sap is a dietary staple (Shaw et al., 2011). The greater ecological impacts of this tactic, as historical examples would suggest, are never fully predictable (Simberloff & Stiling, 1996).

The scuffles with Knotweed turned to battles then to all-out war; in certain jurisdictions, policies against Knotweed were adopted; the annual budget for warring with species designated "invasive" inflated to $21 billion per year (2021) in the United States alone (Fantle-Lepczyk et al., 2022); and Itadori became the poster child for invasive species threatening growth around the world (Rai & Singh, 2000). The war continues to this day and is nearing its centennial anniversary. In affluent countries outside Knotweed's range, season in and season out, governments, conservation nonprofits, and private citizens spray glyphosate and attempt to dig out Knotweed's root systems with tractors; meanwhile, Knotweed simultaneously take advantage of increasing suitable habitat niches perpetually opened through extractive interactions with land. So, while polluted waters, relentless soil disturbance, and climate change–powered storms plant more Knotweed, many continue the self-defeating cyclical war on this plant. To put in other words, Knotweed keep growing because of perpetual anthropogenic changes to the land, are blamed for growing well even though the plant's success is but a symptom not the source of the imbalance, and are here to stay until people change how we live on earth.

As with any long-term war, some human and other-than-human beings adjust to new ways of being, however dark the times may be, and continue their lives. The hope, as after any war, is that eventually a semblance of balance among neighbors will be restored, peaceful language will be remembered, reciprocal ties repaired. This too is quietly taking place in Knotweed's new homelands. Some populations of Knotweed are once again regrowing respectful relations with people. Knotweed offer numerous gifts to land and people with whom the plant did not coevolve.

Knotweed, as ever, provide food. In mid-spring, the plants send up shoots reminiscent of asparagus in shape and texture and earthy rhubarb in taste, which contain protein, vitamins A and C, potassium, zinc, phosphorus, and manganese. People from varied communities across Turtle Island (and Europe) are now familiar with Knotweed and gather the shoots as a seasonal wild food. Despite Knotweed frequently abiding near polluted waterways or being sprayed with herbicide, rendering this potential source of nutritionally dense food detrimental to health, the abundance of Knotweed populations still make it possible for large numbers of people across vast geographies to access this food outside capitalist economies. Knotweed in their opportunistic ways have also entered the marketplace and show up at farm stands across the plant's expanded range. Agricultural soil disturbance regimes and ample available water mean farms commonly host Knotweed populations. Knotweed, in turn, sometimes support produce farmers as a shoulder-season vegetable, rounding out scant selections of early spring cultivated annuals. Knotweed have even entered the fine-dining scene and can now be found on the menus of accoladed restaurants. In parts of northeastern Europe, meanwhile, the large, smooth leaves of Knotweed have entered into the pantheon of wild leaves used for wrapping sarma (Dogan et al., 2015; Łuczaj et al., 2015).

Knotweed, as ever, offer medicine. Over the past decade or so, Knotweed have gained fame as a potent source of resveratrol, an anti-inflammatory antioxidant also found in grape skins, which is used in treatment and prevention of a plethora of diseases, including diabetes and certain cancers (Cucu et al., 2021). In their new homelands on Turtle Island, Knotweed greatly overlap in their distribution range with areas where the tick-borne Lyme disease is most concentrated. Many practitioners of herbal medicine have turned to Knotweed to support patients infected with Lyme, and in preliminary in vitro studies, Knotweed were among the most potent inhibitors of Lyme disease tested (Feng et al., 2020). Knotweed's ability to remove symptoms of this complex infection has been said to be so effective that the connection between the two is seen by some as an auspiciously mystical pairing: a modern remedy for a modern disease, showing up at the source of the problem when most needed.

Knotweed, who bloom at the very end of the growing season in the northern half of their newfound range, also offer nourishment to pollinators, who buzz in great number to the masses of Knotweed flowers for late season meals, at a time when most other flowers have finished offering nectar and pollen (Andros & Apiaries, 2000). Plus, Knotweed often grow in highly anthropogenically modified soils where few other species of flowering plants are able to survive, but hungry pollinators pass through. Keepers of European honeybees on Turtle Island have come to appreciate the honey that bees make during Knotweed blooming time and jars of Knotweed honey are now sold from a few apiaries. The taste is similar to Buckwheat honey (Knotweed's relative), but lighter in flavor and color. Additionally, dried Knotweed stalks have gained favor as an excellent and abundant material for "bee hotels," human-constructed habitats for native mason bees.

Knotweed's robust aboveground growth has led farmers and gardeners in Knotweed-heavy areas to chop stalks of Knotweed at the end of the growing season, chip them, and apply as readily available, extremely renewable, and nutritionally dense mulch. Meanwhile, whole stems added to compost piles aid in compost aeration, given their slow rate of decomposition and hollow centers. Some agrarians also turn to Knotweed for a nutritious and protective foliar fertilizer for their crops (Cvejić et al., 2021) and turn to *Reynoutria sachalinensis* in particular for their fungicidal properties (Herger & Klingauf, 1990).

Knotweed's abundant speed of growth and ability to regrow numerous times in one season after cutting also make Knotweed an appealing source for other plant-based material needs. Knotweed foliage has become a popular source for biomass fuel (Callaghan et al., 1985; Bernik & Zver, 2006), for making paper, and as a nutritionally dense supplemental fodder source for livestock (Kovářová, 2022). Knotweed are also a powerful phytoremediator, with the ability to soak up and hold excess nitrogen from polluted waters and to extract heavy metals from the soil (Nishizono et al., 1989; Hulina & Đumija, 1999). Because the plant can also tolerate growing in highly polluted locales, Knotweed are being investigated as allies in pollution clean-up (Beerling, 1991).

Knotweed's prolific growth in unfamiliar soils, the plant's so-called invasive nature, has invited some people to creatively step into relationship with this plant, to envision Knotweed as a generous resource for a diversity of needs. For some, the motivation for harvesting this plant is also coupled with an eye to slowing Knotweed's growth and repairing ecosystem balance.

3. Re-storyation

When Itadori were taken from their homelands, the plant's stories were left behind. Their rich relationships as members of an intact ecological community were of no interest to the global plant traders. Planted in unfamiliar soils, Knotweed became an object—an adornment for the garden at first;

later, a landscape fright—rather than a being with ecological responsibilities and cultural roles.

Dominant, contemporary stories focus on the negative and costly impacts of Itadori and present restoration through eradication of Knotweed out of a given habitat as the only safeguard against biodiversity loss and decline in socioeconomic wellbeing. Not only do such stories blame Itadori for their ability to grow in places made inhospitable to other plants through repeated anthropogenic actions on the land, but they leave no space to envision the possibility of a bioculturally thriving future with Knotweed in it. Meanwhile, Knotweed have proven to be largely ineradicable in disturbed soils or without a long-term, active commitment to land care. The most violent chemical and physical measures taken against Itadori seem at best to only slow their growth. In other words, restoring to a pre-Knotweed period in the landscape is a fictitious, albeit hopeful, story. Yet, even if Knotweed were to suddenly disappear out of their adopted landscape, restoring the land back to a pre-colonial time is equally unrealistic, given how tangled the composition of plants, animals, diseases, hydrology, and soils has become over recent centuries.

This brief re-story-ative biography of Itadori is a small counteroffering to balance the loudest stories about Itadori growing in Europe or on Turtle Island, which all begin at the chapter when Itadori were deemed "invasive" within certain political jurisdictions. Re-storyation is a practice brought forth by ethnobotanist Dr. Gary Paul Nabhan. "To restore any place," Dr. Nabhan writes, "we must also begin to re-story it, to make it the lesson of our legends, festivals, and seasonal rites. Story is the way we encode deep-seated values within our culture. . . . By replenishing the land with our stories, we let the wild voices around us guide the restoration work we do. The stories will outlast us" (Nabhan, 1997).

For the foreseeable future, Itadori are firmly planted across disturbed soils of Turtle Island and Europe. So, what can restoration look like when the dominant interactions between people and land result in new plantings of Knotweed? Re-story-ation may be the effective restoration measure in this circumstance. For many people across Turtle Island, Northern Europe, and other places outside Itadori's first homelands, Itadori are now the common neighbor to all. What is there to do but to get to know our neighbors' stories and learn how to coexist? If Itadori endure and stories outlast us; what enduring stories can Knotweed carry into the future? It's clear that whichever novel stories we choose to tell ourselves to guide us toward a just future must center reciprocal livelihoods rather than a hierarchy of needs with humans at the top.

Perhaps ecological restoration is less a return to a past time, but a remembering of the stories of responsibility between people and land and a re-storying of biocultural connections between people and the ecosystems we live within. In plant ecologist Dr. Robin Wall Kimmerer's reflection of Dr. Nabhan's words, she writes of a need "to tell ourselves a different story about our relationship to place" (Kimmerer, 2013). It's hard to imagine that the story of categorizing the other-than-human world into a binary of good and bad, then applying violent tactics to eliminate the bad species—all during a time of great biodiversity loss—will hold much wisdom for the generations yet to come. "Our relationship with the land cannot heal until we hear its stories," adds Kimmerer (2013). When it comes to Itadori and other beings whose ancestral stories have been largely silenced out of their biographies and replaced with a bad reputation, we could start by telling stories that aim to see Itadori as beings in their own right, rather than a species whose identity is solely formed by opposition to human desires.

In habitats that are altered beyond recognition, there is a need to have time and space to mourn the loss of species who can no longer live there, but after grief's edges soften, there is a need to learn to celebrate those species who choose to live alongside us in this particular moment in time. What happens when we listen to the stories of people who have found how to relate to invasive plants in culturally and ecologically supportive ways? What happens when we remember that vilified plants are embedded in cultures of people somewhere, that even Japanese Knotweed have their people and their shared stories? People with deep lived-relationships with so-called invasive Itadori can tell stories that offer a bridge to understanding positive ecological and cultural roles Knotweed may take on in a novel range.

Among plant scientists, a new perspective on Invasion Biology has been gaining volume over the past decade. In a time of great extinction and great uncertainty, plant ecologists including Drs. Mark Davis and Matthew Chew suggest that a more restorative categorization of species (at least in highly anthropogenically modified habitats) is based on their function, rather than their origin. "Today's management approaches must recognize that the natural systems of the past are changing forever thanks to drivers such as climate change, nitrogen eutrophication, increased urbanization and other land-use changes. It is time for scientists, land managers and policy-makers to ditch this preoccupation with the native–alien dichotomy and embrace more dynamic and pragmatic approaches to the conservation and management of species—approaches better suited to our fast-changing planet," the authors write (Davis et al., 2011). With this change in perspective, perhaps some day there will be a story to tell of Itadori, with their talent for breaking through concrete and extracting pollutants from soil and water, as accomplices in restoring balance to ecosystems and the people who live in them.

For now and into the foreseeable future, every spring, up from warming soils of the temperate north, Itadori will emerge again. Within a blink, the plant's tiny coral nubs will turn to speckled, hollow shoots, which will turn into impassable, brittle thickets. Knotweed are many things to many beings—in complicated relationship, just like the rest of us. Knotweed stitch the soil back in places where other plants are no longer able; relieve waters overburdened with nutrients; speed the return of concrete back to its most elementary state; feed pollinators, humans, deer; provide medicine for the sick; crumble into durable mulch. Knotweed are gardeners with very strong opinions; world travelers; and keepers of stories about great loss, suffering, anger, celebratory tradition, generous caretaking, and many other subjects.

Literature Cited

Andreotti, G., S. Koutros, J. N. Hofmann, D. P. Sandler, J. H. Lubin, C. F. Lynch, C. C. Lerro, et al. 2018. "Glyphosate use and cancer incidence in the agricultural health study." *Journal of the National Cancer Institute* 110: 509–516.

Andros, C. F. & L. Apiaries. 2000. "*Polygonum cuspidatum* (Japanese knotweed, Mexican bamboo)." *American Bee Journal* 140: 854.

Baker, R. 1988. "Mechanical control of Japanese knotweed in an S.S.S.I." *Aspects of Applied Biology* 16: 189–192.

Barney, J. N., N. Tharayil, A. DiTommaso & P. C. Bhowmik. 2006. "The biology of invasive alien plants in Canada. 5. *Polygonum cuspidatum* Sieb. & Zucc. (*Fallopia japonica* [Houtt.] Ronse Decr.)." *Canadian Journal of Plant Science* 86: 887–905.

Beerling, D. J. 1991. "The effect of Riparian land use on the occurrence and abundance of Japanese knotweed *Reynoutria japonica* on selected rivers in South Wales." *Biological Conservation* 55: 329–337.

Bernik, R. & A. Zver. 2006. "Plant as renewable energy source (RES)." *Acta Agriculturae Slovenica* 87: 355–364.

Butler, B. 1994. *Medical Plants Used by the Ainu*. Master's Thesis, University of Victoria, Victoria.

Callaghan, T. V., G. J. Lawson, A. M. Manwairing & R. Scott. 1985. "The effect of nutrient application on plant and soil nutrient content in relation to biomass harvesting." Pp. 448–453 *in* W. Palz, J. Coombs & D. O. Hall (editors), *Energy from the Biomass: Third E.C. Conference*. Taylor & Francis, Elsevier, Essex & New York.

Child, L. E. & M. Wade. 2000. *The Japanese Knotweed Manual: The Management and Control of an Invasive Alien Weed*. Liverpool University Press, Liverpool.

Chiri, M. 1953. *Bunrui Ainugo Jiten*. Nihon Jomin Bunka Kenkyu, Tokyo.

Conolly, A. P. 1977. "The distribution and history in the British Isles of some alien species of *Polygonum* and *Reynoutria*." *Watsonia* 11: 291–311.

Crawford, G. & M. Yoshizaki. 1987. "Ainu ancestors and prehistoric Asian agriculture." *Journal of Archaeological Science* 14: 201–213.

Cucu, A., G. Baci, Ş. Dezsi, M. Nap, F. I. Beteg, V. Bonta, O. Bobiş, E. Caprio & D. S. Dezmirean. 2021. "New approaches on Japanese knotweed (*Fallopia japonica*) bioactive compounds and their potential of pharmacological and beekeeping activities: Challenges and future directions." *Plants* 10: 2621. https://doi.org/10.3390/plants10122621

Cvejić, R., S. Klages, M. Pintar, L. Resman, A. Slatnar & R. Mihelič. 2021. "Invasive plants in support of urban farming: Fermentation-based organic fertilizer from Japanese knotweed." *Agronomy* (Basel) 11: 1232.

Dauer, J. T. & E. Jongejans. 2013. "Elucidating the population dynamics of Japanese knotweed using integral projection models." *PLOS ONE* 8: e75181.

Davis, M. M., M. M. Chew, R. J. Hobbs, A. E. Lugo, J. J. Ewel, G. J. Vermeij, J. H. Brown, et al. 2011. "Don't judge species on their origins." *Nature* 474: 153–154.

Del Tredici, P. 2017. "The introduction of Japanese plants into North America." *Botanical Review* 83: 215–252.

Dogan, Y., A. Nedelcheva, Ł. Łuczaj, C. Drăgulescu, G. Stefkov, A. Maglajlić, J. Ferrier, et al. 2015. "Of the importance of a leaf: The ethnobotany of sarma in Turkey and the Balkans." *Journal of Ethnobiology and Ethnomedicine* 11: 26.

Earle, M. T. (appendix by C. Lytton). 1897. *Pot-pourri from a Surrey Garden*. Smith, Elder & Co., London.

Fantle-Lepczyk, J. E., P. J. Haubrock, A. M. Kramer, R. N. Cuthbert, A. J. Tubelin, R. Crystal-Ornelas, C. Diagne & F. Courchamp. 2022. "Economic costs of biological invasions in the United States." *Science of the Total Environment* 806: 151318.

Feng, J., J. Leone, S. Schweig & Y. Zhang. 2020. "Evaluation of natural and botanical medicines for activity against growing and non-growing forms of *B. burgdorferi*." *Frontiers in Medicine* 7: 6.

Fennell, M., M. Wade & K. L. Bacon. 2018. "Japanese knotweed (*Fallopia japonica*): An analysis of capacity to cause structural damage (compared to other plants) and typical rhizome extension." *PeerJ* 6: e5246.

Gilbert, O. L. 1992. "The ecology of an urban river." *British Wildlife* 3: 129–136.

Havens, T. R. H. 2020. *Land of Plants in Motion: Japanese Botany and the World*. University of Hawai'i Press, Honolulu.

Herger, G. & F. Klingauf. 1990. "Control of powdery mildew fungi with extracts of the giant knotweed, *Reynoutria sachalinensis* (Polygonaceae)." *Mededelingen van de Faculteit Landbouwwettenschappen Rijksuniversiteit Gent* 55: 1007–1014.

Hollingsworth, M. L. & J. P. Bailey. 2000. "Evidence for massive clonal growth in the invasive weed *Fallopia japonica* (Japanese Knotweed)." *Botanical Journal of the Linnean Society* 133: 463–472.

Hulina, N. & L. Đumija. 1999. "Ability of *Reynoutria japonica* Houtt. (Polygonaceae) to accumulate heavy metals." *Periodicum Biologorum* 101: 233–235.

Indrawan, M., M. Yabe, H. Nomura & R. D. Harrison. 2014. "Deconstructing satoyama—The socio-ecological landscape in Japan." *Ecological Engineering* 64: 77–84.

Kabat, T. J., G. B. Stewart & A. Pullin. 2006. "Are Japanese Knotweed (*Fallopia japonica*) control and eradication interventions effective?" CEE review 05-015 (Systematic Review 21). Collaboration for Environmental Evidence.

Kayano, S. 1996. *Ainu-Japanese Dictionary*. Sanseido, Tokyo.

Kimmerer, R. W. 2013. *Braiding Sweetgrass. Indigenous Wisdom, Scientific Knowledge and the Teachings of Plants*. Milkweed Editions, Minneapolis.

Kimura, Y. & H. Okuda. 2001. "Resveratrol isolated from *Polygonum cuspidatum* root prevents tumor growth and metastasis to lung and tumor-induced neovascularization in Lewis lung carcinoma-bearing mice." *Journal of Nutrition* 131: 1844–1849.

Kobayashi, T., edited by S. Kaner with O. Nakamura. 2004. *Jomon Reflections: Forager Life and Culture in the Prehistoric Japanese Archipelago*. Oxbow Books, Oxford.

Kovářová, M., P. Maděra, T. Frantík, J. Novák & Š. Vencl. 2022. "Effects of knotweed-enriched feed on the blood characteristics and fitness of horses." *Agriculture* (Basel) 12: 109.

Leipe, C., E. A. Sergusheva, S. Müller, R. N. Spengler III, T. Goslar, H. Kato, M. Wagner, A. W. Weber & P. E. Tarasov. 2017. "Barley (*Hordeum vulgare*) in the Okhotsk culture (5th–10th century AD) of northern Japan and the role of cultivated plants in hunter–gatherer economies." *PLOS ONE* 12: e0174397.

Liu, Y., S. Ahmed, B. Liu, Z. Guo, W. Huang, X. Wu, S. Li, J. Zhou, Q. Lei & C. Long. 2014. "Ethnobotany of dye plants in Dong communities of China." *Journal of Ethnobiology and Ethnomedicine* 10: 23.

Locandro, R. R. 1978. "Weed watch. Japanese bamboo, 1978." *Weeds Today* 9(4): 21–22.

Łuczaj, Ł., K. Stawarczyk, T. Kosiek, M. Pietras & A. Kujawa. 2015. "Wild food plants and fungi used by Ukrainians in the western part of the Maramureş region in Romania." *Acta Societatis Botanicorum Poloniae* 84: 339–346.

Mitsuhashi, H. 1976. "Medicinal plants of the Ainu." *Economic Botany* 30: 209–217.

Nabhan, G. P. 1997. *Cultures of Habitat: On Nature, Culture, and Story*. Counterpoint, Washington, D.C.

Nishizono, H., K. Kubota, S. Suzuki & F. Ishii. 1989. "Accumulation of heavy metals in cell walls of *Polygonum cuspidatum* roots from metalliferous habitats." *Plant and Cell Physiology* 30: 595–598.

Ohwi, J. 1984. *Flora of Japan*. Smithsonian Institution, Washington, D.C.

Peng, W., R. Qin, X. Li & H. Zhou. 2013. "Botany, phytochemistry, pharmacology, and potential application of *Polygonum cuspidatum* Sieb.et Zucc.: A review." *Journal of Ethnopharmacology* 148: 729–745.

Rai, P. & J. A. Singh. 2020. "Invasive alien plant species: Their impact on environment, ecosystem services and human health." *Ecological Indicators* 111: 106020.

Robinson, W. 1870. *The Wild Garden*. J. Murray, London.

Seiger, L. A. & H. C. Merchant. 1997. "Mechanical control of Japanese knotweed (*Fallopia japonica* [Houtt.] Ronse Decraene): Effects of cutting regime on rhizomatous reserves." *Natural Areas Journal* 17: 341–345.

Shaw, D. 2013. "*Fallopia japonica* (Japanese knotweed)." CABI Compendium. https://doi.org/10.1079/cabicompendium.23875

Shaw, R. A., R. I. Tanner, D. Djeddour & G. Cortat. 2011. "Classical biological control of *Fallopia japonica* in the United Kingdom—Lessons for Europe." *Weed Research* 51: 552–558.

Shimoda, M. & N. Yamasaki. 2016. "*Fallopia japonica* (Japanese Knotweed) in Japan: Why is it not a pest for Japanese people?" *Geobotany Studies* 447–473.

Simberloff, D. & P. Stiling. 1996. "Risks of species introduced for biological control." *Biological Conservation* 78: 185–192.

Tan, S., G. Li, Z. Liu, H. Wang, X. Guo & B. Xu. 2022. "Effects of glyphosate exposure on honeybees." *Environmental Toxicology and Pharmacology* 90: 103792.

von Ehrenstein, O. S., C. Ling, X. Cui, M. Cockburn, A. S. Park, F. Yu, J. Wu & B. Ritz. 2019. "Prenatal and infant exposure to ambient pesticides and autism spectrum disorder in children: Population based case-control study." *BMJ* 364: I962

Williams, D. 2017. *Ainu Ethnobiology*. Society of Ethnobiology, Tacoma. https://ethnobiology.org/sites/default/files/publications/contributions/ainu-ethnobiology-final-web-01-09-2018.pdf

Wood, J. 1884. *Hardy Perennials and Old-fashioned Garden Flowers: Describing the Most Desirable Plants for Borders, Rockeries, and Shrubberies, and Including Both Foliage and Flowering Plants*. L. U. Gill, London.

Yoshioka, K. 1974. "Volcanic vegetation." Pp. 237–267 *in* M. Numata (editor), *The Flora and Vegetation of Japan*. Kodansha, Tokyo.

Zhang, H., C. M. Li, S. T. Kwok, Q. Zhang & S. W. Chan. 2013. "A review of the pharmacological effects of the dried root of *Polygonum cuspidatum* (Hu Zhang) and its constituents." *Evidence-based Complementary and Alternative Medicine* 2013: 1–13.

CHAPTER 4

Cooking with Invasive Species: Culinary Suggestions for Promoting Control through Consumption

Alana N. Seaman & Alexia Franzidis

A GLASS OF WINE. A lean, perfectly charred steak with a rich earthy seasoning, topped with a local dark summer berry compote reduction, and accompanied by a mashed root vegetable medley and a pleasantly bitter salad of fresh harvested local greens tossed in a light, sweet, tangy, floral vinaigrette. Elegant, understated. And . . . with the right ingredients, excellent for the environment!

Control through consumption, or the harvesting of "invasive alien species" (IAS) for humans to eat, is not only delicious, it also represents a realistic means of supporting an important environmental cause on a personal level. It is considered a viable option in curbing the proliferation of many non-native plants and animals, and research suggests that people are generally supportive of IAS management and would like to do their part toward conservation efforts given opportunities to do so (Wald et al., 2019). However, while popular resources such as *Southern Living* and *Garden & Gun* magazines are encouraging audiences to harvest various IAS for use in specific recipes such as kudzu lemonade (Hayes, 2019), dandelion greens salad, and Himalayan blackberry shortcakes, sorbets, and syrups (Arrington, 2016), little has been written about the topic, particularly in the academic literature.

Nonetheless, scholars contend that, if framed correctly, almost any product can become an accepted and even sought-after food item (Gyimothy & Mykletun, 2009; Seaman et al., 2019, 2021). Thus, if the consumer is to play a role in control, information needs to be available to communicate what each species tastes like and how it could be prepared and consumed.

This chapter aims to do just that! It provides an overview of the most prevalent IAS in the United States and their potential culinary applications. IAS are grouped according to three categories: Plants; Seafood/Aquatic Animals; and Land Creatures. For each IAS, there is information about where in the U.S. the species has spread, which region it is native to, how it tastes, and how it can be used. IAS are listed alphabetically in each category.

The information in this chapter was inspired by photos and accompanying comments about "control through consumption" and "eating invasives" posted publicly on popular social media platforms. Researchers often utilize these sources to get a glimpse into a social world or phenomena that might otherwise be hard to tap into. Many control through consumption efforts are taking place at a grassroots level and are therefore difficult to identify and study. However, the individuals who are currently foraging, preparing, and consuming IAS within the U.S. and sharing information about their experiences online represent an important resource for this emerging topic. Thus, their collective knowledge, advice, and thoughts on edible invasive species were used to generate the information provided.

Background information (e.g., notable characteristics, prevalence, toxicity, etc.) was verified/gathered from official sources such as the National Invasive Species Information Center run by the U.S. Department of Agriculture (USDA). Please note that only brief descriptions of each species' conspicuous characters are given. Plants are not described in enough detail to distinguish them from all other plants, some of which may be inedible or toxic. If you are not already familiar with the plant, consult more authoritative botanical references and make very sure you have correctly identified it before foraging!

Plants

Autumn Olive

Also known as the Japanese silverberry. Shrub up to 20 feet high and 30 feet wide. Dense branches. Small round smooth berry, silvery with brown when young, ripens to red. Located in grasslands, fields, open woodlands, and other disturbed areas.

Prevalence in the U.S.: Found in states between Maine and Virginia (south), and west to Wisconsin. Many U.S. states; however, it is dominant in Central and Midwestern states.

Native to: Eastern Asia between the Himalayas and Japan.

Flavor profile: Sweet and sour. Similar to tart cherries or pomegranates. Can be tart before berry is fully ripened.

Suggested use: Preserve in honey and flavored syrup to use over yogurt or granola. Puree and dry for delicious fruit leather. Also makes a good jam. Made into a sauce, pairs well with chocolate desserts.

Black Locust

Medium-sized hardwood deciduous tree 30–50 feet in height and 20–35 feet wide. Very upright tree with narrow crown and long branched thorns on trunk. Dull green leaves with a pale underside. Hanging clusters of loosely drooping clumps of pale yellow to white fragrant flowers. Produces pealike pods, which are *poisonous*.

Prevalence in the U.S.: Continental U.S.

Native to: Eastern U.S. states and Appalachian and Ozark Mountains.

Flavor profile: Flowers have a sweet pea honey flavor. Bark, leaves, and pods are toxic.

Suggested use: Flowers can be used for teas and syrups.

Burdock

Tall flowering stem with pink-purple flower clusters. Knee to shoulder height. Large wavy-edged, heart-shaped leaves with a woolly whitish underside. Flower head has burrlike appearance similar to a thistle; clinging hooks inspired Velcro.

Prevalence in the U.S.: All U.S. states except Florida and Hawai'i.

Native to: Northern Europe.

Flavor profile: Similar to lotus root. Earthy, nutty, and sweet/bitter like an artichoke. Crunchy texture like a radish.

Suggested use: Use roots only. These can be roasted or braised. Excellent with Asian-inspired seasonings. Avoid young burdock greens as these are extremely bitter.

Chickweed

A weed common in U.S. lawns and golf courses. An annual in colder climates, evergreen in warmer climates. Low growing and dense, 4–6 inches in height. Bright green smooth leaves. Small five-petaled white flowers.

Prevalence in the U.S.: Throughout most U.S. states except in the Southeast.

Native to: Europe and Asia.

Flavor profile: Fresh grassy flavor akin to cornsilk.

Suggested use: Can be used like sprouts. Leaves are a nice addition to salads. Leaves, flowers, and stems can be used in various baked or roasted dishes, or added to soups and stews.

Dame's Rocket

Plant is 2–4 feet tall with a tall stem and clusters of small four-petaled flowers at the top. Pointed lance-shaped leaves 2–6 inches long. Blooms in spring. Can be white to purple. A relative of garlic mustard, it often grows alongside that plant.

Prevalence in the U.S.: Northern U.S. states.

Native to: Mediterranean region and central Asia.

Flavor profile: Entire plant is edible in moderation. Mustardy, spicy, but mostly mild.

Suggested use: Buds can be lightly sautéed like fresh greens. Young leaves can be used in salads.

Dandelion

Dark green leafy weed, ubiquitous on American lawns. Has small yellow composite flowers.

Prevalence in the U.S.: Found throughout the U.S.

Native to: Europe and Eurasia.

Flavor profile: Leaves and flowers have a mild earthy flavor. Leaves are pleasantly bitter with a texture akin to arugula but more tender than kale.

Suggested use: Leaves for fresh greens/salads and for teas with supposed medicinal use. Flower can be used to make wine. Ground roasted root is a coffee substitute.

Daylily

Colorful yellow, orange, and red flower with a tall stem, 3 feet tall.

Prevalence in the U.S.: Found in all regions in the U.S.

Native to: Asia.

Flavor profile: Buds and flowers are edible and have a sweet/spicy or peppery flavor. Buds are green, with a flavor akin to radish with green beans. Stalks have a texture like lemongrass although not a lemon flavor. The tuber is like a potato, similar in taste to raw sweet potato; it is similar to but sweeter than jicama.

Suggested use: Flowers can be stuffed with sweet or savory fillings. (A few people are allergic or find they have a laxative effect!) Buds can be steamed, boiled, roasted, grilled, sautéed, or even pickled. A savory jam made from flowers pairs well with creamy cheeses. Tubers can be boiled or steamed then sautéed.

Garlic Mustard

Undergrowth and at forest edges. Can be difficult to identify, but smells of garlic when leaves are crushed. Dark green kidney-shaped leaves with scalloped edges. Mature leaves are more triangular and longer with sharper teeth. Plants mature to 3.5 feet tall. Flowers produce four small white petals, bloom in early spring and die back by summer.

Prevalence in the U.S.: Northeastern, Midwestern, and Northwestern states.

Native to: Europe, western and central Asia, Northwestern Africa, Iberia, the British Isles, and northern Scandinavia.

Flavor profile: Leaves taste like a cross between bitter mustard greens and garlic. Flower buds have a horseradish-like taste. Young and shade plants are less bitter. Contains very small amounts of cyanide, so some suggest using older plants only after boiling to remove it. In fact, the young leaves of seedling plants have highest concentrations (Cipollini & Gruner, 2007). No illnesses have been reported among pesto consumers, so use of raw adult leaves in moderate quantity is likely safe in practice.

Suggested use: Adds spice to dips, sauces, salads, and stir-fries. Excellent as a pesto base.

Himalayan Balsam Flower

Plant 3.5–6.5 feet in height. Has broad dark green leaves with red veins and red tinge at edges. Soft green to red stems. Flowers are pink with a hooded shape. Seed pods have a kite shape and green and red veins.

Prevalence in the U.S.: Northern states along the East Coast, the West Coast including Montana and Idaho.

Native to: Himalayan mountains.

Flavor profile: Leaves are lightly bitter yet mild. Seeds have a nutty flavor.

Suggested use: Leaves can be used as fresh greens. Flowers can be used for jams and teas. Flowers can also be fermented for a vinegar or kombucha drink.

Himalayan Blackberry

Woody shrub up to 13 feet tall with stout stem and stiff sharp hook-shaped prickles. Leaves alternate with five leaflets. Leaves have toothed edges. Flowers are clustered, five-petaled, white, about 1 inch across. Berries ripen in late summer or early fall to a black and dark purple color.

Prevalence in the U.S.: Pacific Northwest and California.

Native to: Armenia and Northern Iran.

Flavor profile: Less flavorful than typical blackberry, bland although sweet and juicy.

Suggested use: Like other berries in pies, jams, and compotes.

Honeysuckle

Aggressive vine with no tendrils, hairy when young. Leaves opposite, egg-shaped, usually persisting into or through winter. Flowers in pairs in leaf axils, with white spreading lips, pleasant scent.

Prevalence in the U.S.: Most Eastern states in the U.S.

Native to: Northern latitudes in North America and Eurasia.

Flavor profile: Flowers and nectar have a light floral honey flavor. May have notes of apple, fresh greens, and/or citrus.

Suggested use: Tea, vinegar, beer, kombucha, and syrups. Berries are *not* edible.

Japanese Knotweed

Stalks like thick asparagus. New shoots are red to purple. Leaves unroll as plant turns green. Bamboo-like canes grow up to 10 feet high.

Prevalence in the U.S.: Found in 42 states coast to coast. Not found in the Deep South Gulf States, states within the arid Southwest, or the highest parts of the Rocky Mountains.

Native to: East Asia, specifically Japan, China, and Korea.

Flavor profile: Similar to rhubarb, or a cross between asparagus and rhubarb. Tart, crunchy, juicy, tangy, and earthy.

Suggested use: Shoots can be eaten raw, or used for a cocktail flavoring, or in place of rhubarb in a pie. Can also be pickled and used for various sweet/tangy recipes. Leaves and young shoots can be quick-pickled.

Kudzu

Dark leafy green vine with hairy stem, large tuber-producing taproot, and alternating leaves 2–8 inches long with three leaflets. Leaflets broadly oval, often lobed. Stems are yellow green with dense erect gold hairs. Flowers usually red to purple, rarely white, fruit a legume.

Prevalence in the U.S.: Southeastern states in the U.S.

Native to: China, Japan, and Indian subcontinent.

Flavor profile: Vine tips are snow pea–flavored. Fresh leaves have a spinach-like flavor. Flowers have a fruity taste.

Suggested use: Young tender leaves can be used like collards or spinach for quiche. Root starch, called kuzu in Japan, can be a thickening agent like cornstarch. Grape-smelling blossoms can be used for jelly, syrups, and wines.

Lamb's Quarters

Herb can be 2–6 feet tall. Leaves are somewhat diamond-shaped, toothed, light green on top with whitish underside. Flowers small, greenish, in dense clusters, each producing a small black seed.

Prevalence in the U.S.: All U.S states

Native to: Europe.

Flavor profile: Mineral-rich green tasting akin to chard or spinach.

Suggested use: Often used like other greens in Indian dishes. Nice addition to fritters or soups. Should be cooked if eaten in more than small amounts;

raw greens contain oxalic acid (like spinach does) and may accumulate nitrates from the soil. Seed can be washed and dried to remove bitterness, then included in flours, porridges, etc.

Mallow

Plant 6–24 inches high with pink or white hollyhock-like flowers along long stems with circular wavy-edged leaves. Flowers early spring to mid-fall.

Prevalence in the U.S.: Throughout the U.S.

Native to: Europe, North Africa, and Asia.

Flavor profile: Mild with an almost nonexistent flavor although somewhat earthy. Leaves have a touch of sweetness.

Suggested use: Leaves are mucilaginous and can be used as a thickening agent for soups and stews. Also makes a nice pasta flavoring base. Roasted leaves and flowers can be used for a tea.

Mugwort

Aromatic flowering plant 3 feet high. Smells vaguely like sage. Dark green deeply lobed leaves vaguely reminiscent of parsley. Underside of leaves is silvery white with downy hairs. Produces clusters of 7–10 small yellow to reddish brown flowers.

Prevalence in the U.S.: Eastern states.

Native to: Europe and Asia.

Flavor profile: Slightly bitter and was a popular flavoring for beer before hops was used.

Suggested use: Dried leaves can be used for teas and tinctures. Also a savory flavoring for fat, meat, and fish. Put in food processor with equal amount of salt for a fresh seasoning.

Purple Deadnettle

A groundcover with fuzzy spade-shaped leaves and purple to pink small cone-shaped flowers. Similar to mint in general appearance (related). Grows in lawns and empty lots.

Prevalence in the U.S.: Most U.S. states except Nevada, Arizona, New Mexico, Wyoming, North and South Dakota, and Minnesota.

Native to: Europe and Asia.

Flavor profile: Entire plant is edible with a grassy flavor. Can be slightly minty. Like a cross between spinach and overcooked green beans with a hint of spice and pepper.

Suggested use: Green salads. Can be sautéed. Adds a nice flavor to pesto. Add to dips for a fresh green flavor.

Watercress

An invasive *Nasturtium* species found only in wet habitats. Herb with circular to oval leaflets. Long clusters of white cross-shaped flowers bloom in early spring through early fall. Produces long narrow seedpods.

Prevalence in the U.S.: Found throughout the U.S. except North Dakota.

Native to: Andes mountains of South America.

Flavor profile: Sweet and peppery flavor.

Suggested use: Best used as a solid green garnish or as a cooked green. If grown in contaminated water, it can carry germs and parasites including liver flukes. Harvest only from above the water if possible, then rinse thoroughly, and/or cook in a large amount of salted water.

White Mulberry

Small to medium-sized tree to 40 feet tall, with spreading branches, broadly egg-shaped, toothed, often lobed leaves. Each fruit is a cluster of tiny berries, normally red to purple in color, rarely white. (Note: the red mulberry is native to North America, not invasive.)

Prevalence in the U.S.: Found in every state except Nevada, Alaska, and Hawai'i.

Native to: China.

Flavor profile: Berries have a sweet mild honey-like flavor.

Suggested use: Berries can be used for pie and tart fillings, jellies, jams. Pairs well with meats and cheeses. Ferment into a Bosnian drink called smreka. Also nice as a homemade vinegar. Can be dried for snacks.

Wild Garlic / Field Garlic

Short bulb, stem, and leaf with clusters of small white flowers. Leaves are dark green, pointed, and long. Grows in dense clumps on woodland floor. Harvest in late winter to early spring. Garlic odor important to avoid collecting poisonous monocots.

Prevalence in the U.S.: States in the West Coast and the Eastern half of the continental U.S.

Native to: Europe, Africa, and the Middle East.

Flavor profile: Lighter in flavor than traditional garlic.

Suggested use: Can be used raw or cooked. Works well for pesto and pasta sauces, salads, and soups. Also a lovely addition to a seasoned butter. Can be dehydrated and used as a shelf-stable dry seasoning.

Wineberry

An Asian species of raspberry. Multi-stemmed, spiny, dense shrub with thorny stems with reddish hairs, leaves with three leaflets, whitish underneath. Flowers with white petals. Fruits resemble commercial raspberries.

Prevalence in the U.S.: Found predominantly in Eastern U.S. states, from southern Georgia to Tennessee, and the Great Lakes to Maine.

Native to: China, Japan, and Korea.

Flavor profile: More tart and juicier than regular raspberries, sticky to the touch.

Suggested use: Use like other tart berries: syrups, desserts, baked goods.

Seafood / Aquatic Animals
Blue Catfish

Freshwater fish. Steeply sloped straightforward profile and head. Pale gray/blue with no spots. Can exceed 100 pounds. Deeply forked tail fin. Anal fin with 30–36 rays.

Prevalence in the U.S.: Considered invasive in Chesapeake Bay.

Native to: The Mississippi, Missouri, Ohio, and Rio Grande River basins.

Flavor profile: Like striped bass.

Suggested use: Can be grilled or pan-fried. For best flavor cut off the red meat, soak in salt water for two days in the refrigerator. Do not overcook—will get rubbery. Great deep-fried and served as a fish taco with a spicy slaw.

Carp

Freshwater fish. Pelvic fin set back on the body with pectoral fins low on the sides. Silvery with large dark blotches on back and sides. Can jump up to 10 feet in the air when disturbed.

Prevalence in the U.S.: Almost every U.S. state.

Native to: Europe and Asia.

Flavor profile: Meat is white, firm, and mild. Absorbs seasoning well. Intramuscular bones throughout.

Suggested use: Best fried, marinated, or used for tacos, wontons, and pot stickers.

Green Crab

Measures 2.5–4 inches long resembling a Dungeness crab. Has five triangular spines behind the eyes and three rounded bumps between the eyes. Dark to olive green. Green crabs do not have hairs like other similar small crab species.

Prevalence in the U.S.: West Coast and Northeast Coast.

Native to: Northern Africa and Europe.

Flavor profile: Sweeter than a blue crab, complex and rich.

Suggested use: Texture however is inferior, akin to processed meat, best used for empanada fillings and flavoring soups.

Lion Fish

Saltwater fish up to 18 inches long with 18 venomous spikes. Brown to maroon in color with white stripes. Has a spiny back and pointed pectoral fins, both of which are spotted.

Prevalence in the U.S.: Southeastern coast between Florida and North Carolina.

Native to: Indo-Pacific.

Flavor profile: Sweet, mild, moist, buttery, and tender fish, flavor somewhere between grouper and mahi. A white fillet that is flaky. Somewhere between shrimp and lobster, it is firm enough for addition to ceviche.

Suggested use: Ceviche. Can be sautéed or pan-fried. Holds up well to low-temperature smoking.

Rusty Crayfish

Appears similar to other crayfish but with a circular rust-red patch on either side just before the tail; otherwise greenish gray to reddish brown. Large by comparison to other crayfish, 3–5 inches.

Prevalence in the U.S.: Eastern U.S.

Native to: The Ohio River Basin in parts of Ohio, Kentucky, and Indiana.

Flavor profile: Lobster flavor but more subtle.

Suggested use: In a crawfish boil. Can be used in gumbo, étouffée, and po-boys.

Samoan Crab / Mud Crab

Recognized by their broadly flattened back legs with paddle-like last segments. They have a smooth carapace with very robust claws. Deep green in color (Rowling et al., 2010).

Prevalence in the U.S.: Hawai'i and Florida.

Native to: Eurasia.

Flavor profile: Sweet and moist meat.

Suggested use: Nice filling for pastas or to add seafood flavor to soups and stews. In a salad, pairs well with citrus.

Snakehead

Elongated predatory fish that looks like a snake. Dark brown to tan with irregular dark blotches. Has a long dorsal fin, long mouth, and shiny teeth. Can breathe air with gills.

Prevalence in the U.S.: Predominantly in Florida.

Native to: Parts of Africa and Asia.

Flavor profile: Mild, firm, and flaky.

Suggested use: Can be beer-battered, fried, pan-fried, or grilled. Remove skin and fat. The younger the better for best flavor.

Tench

Large minnow. Freshwater/brackish water fish. Looks like a cross between a smallmouth bass and a carp. Up to 27 inches long with a record weight of 15 pounds. Large bodied, stout, and small-scaled. Brown in color with a bronze belly.

Prevalence in the U.S.: Scattered throughout freshwaters in the U.S.

Native to: Eurasia.

Flavor profile: Bony. Can taste muddy and should be harvested from clear water.

Suggested use: Best fried or cured with a sweet-and-sour marinade.

Land Creatures

Green Iguana

Usually green but can also be somewhere between brown and black. Has a row of spikes behind the center of the neck that goes down the back and tail. Can grow up to 5 feet long but only about 7–17 pounds.

Prevalence in the U.S.: Florida, Texas, and Hawaiʻi.

Native to: Several Caribbean islands, Central America, and warm regions of South America.

Flavor profile: White meat akin to chicken, pork, or grouper. Texture is similar to rabbit. Milder than gator. Lean, can be slightly gamy.

Suggested use: Fry, roast, boil, or deep-fry.

Sika Deer (Similar to Axis Deer)

Medium-sized deer with a furrowed brow. Moves in herds. Has white spots throughout adulthood. Has a shorter snout and smaller ears than a whitetail deer. Antlers grow up to 15 inches.

Prevalence in the U.S.: Texas, Virginia, and Maryland.

Native to: China, Korea, Taiwan, and Japan.

Flavor profile: Strong-flavored dark meat similar to elk. High-quality venison. Sweet, slightly rich and beefy.

Suggested use: Steaks (e.g., loin, ribeye, strips, etc.). Jerky. Ground meat is excellent in chili, stew, soups, and the like.

Wild Boar

Boar with bristled hair, scrunched face, pointed snout, beady eyes, and large ears. Can be white, black, brown, and/or red. Often with tusks. Up to 220 pounds, 2–4 feet high, and 3–6.5 feet long.

Prevalence in the U.S.: In 35 states of the continental U.S. (Southeastern and Western states).

Native to: Eurasia and North Africa.

Flavor profile: Not gamy. Strong nutty and rich flavor likened to a cross between pork and beef.

Suggested use: Pairs well with red wine. Ground meat is an easy substitute for any other ground meat dish (e.g., tacos, dumpling fillings, etc.). Can also be ground and mixed with other red meats for a tasty burger. Processed into sausage, has a nice unique flavor. Holds up well to smoking.

We hope this list provides inspiration for your future culinary adventures. Enjoy! And, as one post noted, "if you can't beat them, eat them"! For more information, check out https://www.invasivespeciesinfo.gov.

Literature Cited

Arrington, D. 2016. "Black, blue, or red—These berries are all delicious." *The Sacramento Bee*, 27 July 2016 (updated). https://www.sacbee.com/food-drink/recipes/article91991317.html

Cipollini, D. & B. Gruner. 2007. "Cyanide in the chemical arsenal of garlic mustard, *Alliaria petiolata*." *Journal of Chemical Ecology* 33(1): 85–94.

Gyimothy, S. & R. J. Mykletun. 2009. "Scary food: Commodifying culinary heritage as meal adventures in tourism." *Journal of Vacation Marketing* 15(3): 259–273.

Hayes, H. 2019. "How to make kudzu lemonade." *Southern Living Magazine*, 19 April 2019 (updated). https://www.yahoo.com/lifestyle/kudzu-lemonade-205544489.html?

Rowling, K., A. Hegarty & M. Ives (editors). 2010. Giant mud crab. Pp. 147–150 *in Status of Fisheries Resources in NSW 2008/09*. Industry & Investment NSW, Cronulla, New South Wales. https://fish.gov.au/Archived-Reports/2012/reports/Documents/Rowling_et_al_2010.pdf

Seaman, A. N., A. Franzidis, H. Samuelson & S. Ivy. 2019. "Eating local invasives: Chefs as central to the control through consumption approach." Presentation at American Association of Geographers Annual Meeting. Denver, Colorado.

Seaman, A. N., A. Franzidis, S. Ivy & H. Samuelson. 2021. "Eating invasives: Chefs as an avenue to control through consumption." *Food, Culture & Society* 25(1): 108–125.

Wald, D. M., K. A. Nelson, A. M. Gawel & H. S. Rogers. 2019. "The role of trust in public attitudes toward invasive species management on Guam: A case study." *Journal of Environmental Management* 229: 133–144.

CHAPTER 5

Can We Love Invasive Species to Death?

Sara E. Kuebbing, Joshua Ulan Galperin & Martin A. Nuñez

In 2010, a new word was born: invasivore. Journalist James Gorman coined the phrase in a news article describing Floridian lionfish derbies and Chicagoan chefs serving "Kentucky tuna." Invasivores are adventurous eaters who dine on non-native, invasive species (Gorman, 2010), which are a species that is not native to that ecosystem whose introduction does or is likely to cause economic or environmental harm or harm to human health (Executive Order 13112, 1999). Since 2010, the word has become common and, perhaps, grown into an ideology in some scientific and haute cuisine circles. There are now prominent websites like Eat the Invaders (http://www.eattheinvaders.org) and dozens of published cookbooks sharing recipes for non-native animals and plants (Box 1). Although there is no scientific support for the effectiveness of invasivory in controlling populations of invasive species (Nuñez et al., 2012; Galperin & Kuebbing, 2013; Pasko & Goldberg, 2014), what is clear is that there is growing societal support and excitement for the idea (Barnes et al., 2014; Seaman et al., 2021).

Support and excitement for invasivory stems from both culinary and conservation arenas. Wild food harvesters and locavores embrace invasivory because it fits within a dietary ethos of eating animals and plants that people can easily forage near their homes (Landers, 2012a, 2012b; Linnekin,

2018; de Lapparent Alvarez, 2022). Chefs also frequently welcome the idea of peculiar and unfamiliar ingredients with a good backstory of environmentalism and sustainable eating (Herzog, 2011; Kleiner, 2019; Parola, 2022; Seaman et al., 2021). Biologists and conservationists enthusiastically stoke the excitement of adventurous eaters for several reasons (Roman, 2004; Morgan, 2015; Illinois Department of Natural Resources, 2018; Florida Fish and Wildlife Conservation Commission, 2022). First, eating invaders and recipe sharing is a fun (and tasty!) way to educate the public about what non-native, invasive species are and how they can negatively impact local native species and ecosystems. Second, local foraging and hunting can be an important nutritional resource and cultural activity for communities in urban and rural locales (McLain et al., 2013). Third, some biologists suggest that eating invasive species may be a beneficial tool for managing non-native populations (Roman, 2004; Morgan, 2015).

Because of the broad base of support, the idea of invasivory (and the fun word play) has gained popularity with journalists and newsrooms. You can find catchy headlines that shout *"if you can't beat 'em, eat 'em," "eradication through mastication,"* and *"gnawing away at invasives"* in local weeklies and international newspapers. These articles showcase a creative, easy, and tasty "win-win" solution that, some argue, could simultaneously reduce environmental harm of non-natives, provide people with delicious food, and maybe even earn some people money (Keller, 2013; Conant, 2020; Hongoltz-Hetling, 2021).

Invasivory activities come in different flavors. For some people, invasivory may be a recreational activity. Recreational invasivores will likely see multiple benefits from harvesting and hunting invasive species. Wild food foraging takes people outside to enjoy the natural world, increases the diversity of ingredients in home kitchens, celebrates familial or community traditions, and can decrease household grocery budgets (depending on the equipment or tools necessary for harvesting). Recreational invasivory may also be an important educational tool for bringing community members together to learn how to recognize specific invasive species. For other people, invasivory may be an economic opportunity. If there are enough hungry invasivores who do not forage invasive species for themselves, then an entrepreneurial harvester may be able to make a profit by supplying the demand for edible invasive species.

Recreational and commercial invasivores likely prioritize different facets of harvesting invaders and these prioritizations may lead to different invasivory outcomes. In particular, profit-motivated commercialized invasivory may lead to perverse incentives that prioritize profit over invasive species management. Although the idea of invasivory as a management tool is popular, some scientists and environmentalists (including the authors of this chapter) raise biological, economic, and legal concerns about the efficacy of human consumption and about unpredictable outcomes (Nuñez et al., 2012; Galperin & Kuebbing, 2013; Pasko & Goldberg, 2014). Importantly, these critiques may only apply to recreational invasivory activities, only economic invasivory activities, or both. We summarize these critiques by answering three key questions: First, is it biologically possible for invasivory to reduce the size of invasive species populations? Second, is invasivory legal? And last, what happens if we love an invasive so much, we end up embracing that invasive? We address whether the critiques are pertinent to recreational versus economic invasivory and suggest, when possible, how recreational or commercial invasivores may reduce the risk of invasivory activities. We use examples of common non-native, invasive species that are popular invasivore targets in the United States, and we limit our discussion of legal restrictions on invasivory movements to American federal and state laws. While our examples may draw from a single nation and may be most pertinent to invasivores in the United States, we suspect that one will find similar examples and caveats in global invasivory campaigns.

Why Invasivory?

The prevailing evidence supporting invasivory as a viable management tool is the fact that humans have already hunted many species into extinction. Globally, overharvesting species for food or other human uses is one of the top five causes of biodiversity decline (Wilcove et al., 1998; IPBES, 2019). In the United States, European colonizers and their descendants have been indicted in the extinction of the passenger pigeon (Halliday, 1980) and the near extinction of the Atlantic cod (Murawski, 2010), American ginseng (McGraw, 2001), wild turkey (Mosby, 1949, 1975), and American buffalo (Taylor, 2007). All these native species were historically prolific in North America, and the subsequent collapse of these populations suggests that human overexploitation—especially commercialized (not recreational) exploitation—is a potent and effective means for exterminating widespread and populous species.

The underlying logic for invasivory as a management tool parallels the logic of another common—and hotly debated—invasion management tool.

Biological control, or "biocontrol," is a practice that traditionally involves introduction of *another* non-native species that is a pest, parasite, or pathogen of the target invasive species. Like invasivory, the goal is for the biocontrol species to eat or infect the target invasive species, which will lead to a reduction in population (Hajek et al., 2016). The history of biocontrol management strategies is long, tracing back to the late 1800s (Hajek et al., 2016) and providing a rich set of examples of the long-term ecological consequences of biocontrol management. Because many examples exist of unintended consequences of biocontrol—the biocontrol agent failing to reduce the population size of the target invader, the biocontrol agent attacking native species and reducing native populations, or other unpredictable effects of the biocontrol agent on interactions among other species in the ecosystem—there is a significant debate about the efficacy and utility of biocontrol as a viable and ethical management strategy (Simberloff, 2012; Seastedt, 2015). While our goal in this chapter is not to review the lengthy history of the impacts of non-human biocontrol agents for invasive species management, we do emphasize the similarity between "invasivory" and "biocontrol" and highlight that these types of approaches can carry large risks and lead to unpredictable and unforeseen outcomes (Simberloff, 2012). Importantly, while the unforeseen outcomes of biocontrol management tend to be ecological, using humans themselves as the biocontrol agent may lead to ecological problems and problems associated with unforeseen human behaviors (e.g., transporting invasive species into new areas for harvesting).

Is It Biologically Possible for Invasivory to Reduce the Size of Invasive Species Populations?

As with many traditional biocontrol efforts, some biologists question whether historic consumption-driven declines in native populations are strong evidence that humans could cause similar declines in non-native populations (Nuñez et al., 2012). In other words, are widespread native and non-native, invasive species biologically comparable? There are many ecological traits that are likely to make a non-native species invasive and differentiate it from native species. Non-native, invasive species tend to be fecund, lack many predators and pathogens, and are good at dispersing to new regions and acquiring limited resources when they arrive (van Kleunen et al., 2010; Liu et al., 2017). This suite of biological traits means that it will generally take larger harvest efforts to reduce a well-established, fast-growing invasive population than it would to reduce a stable, native population. Additionally, the decline of many of these numerous native species' populations was not *just* from human consumption, but also human-induced habitat loss, introduction of new diseases and pathogens, or introduction of other non-native species that *together* led to the precipitous decline of the native populations (Mosby, 1949, 1975; Halliday, 1980; McGraw, 2001; Taylor, 2007). Thus, it is unclear whether human consumption alone could significantly reduce the populations of widespread invasive species.

Because reducing population size of invasive species is not trivial, managers of invasive species should know how harvesting individuals affects the growth rate of the population (Pasko & Goldberg, 2014). Wildlife and fishery biologists frequently use mathematical models to understand the impacts of human harvest on wild populations. These models are derived from long-term data that track individual animals or plants through time to understand birth, death, and survival rates for individuals of different ages (adult or juvenile) or sizes (larger or smaller adults). People can then use these models to predict how harvesting individuals of different ages or sizes affects the overall growth rate of a given population as well as estimate how many individuals of certain ages or sizes need to be removed to make populations shrink through time.

Using these population growth models, invasivores can determine the biological impacts of their diets. Invasivores may be restricted to harvesting specific life stages (such as large adults) or edible, appealing, and marketable portions (such as leaves or fruits). It is possible that our favorite life stage for consumption, or the most delicious part of an invasive plant, may not align with the most appropriate life stage for management. Mismatches between the most appropriate life stage for management and the preferred life stage for consumption could lead to a few important implications that invasivores should consider.

For some species, invasivores may achieve population declines of the tasty invader, but to achieve those declines they may have to consume *more* individuals of the preferred life stage than they would if they consumed other life stages. In the Mississippi River, fishery biologists and chefs are excited about remarketing invasive carp as "Kentucky tuna," or more recently, "Copi" (The Economist, 2022) and serving the fish in four-star restaurants (Pugliese, 2010). Currently, commercial fishers selectively capture larger fish (Weber et al., 2011; Garvey et al., 2015), although fishery biologists have found that the most efficient way to control carp populations is by removing fish of all size groups (Weber et al., 2011; Tsehaye et al., 2013). Biologists estimate that removing 33% of medium-sized carp would be as effective for reducing carp populations as removing 75% of the largest-sized carp (> 50 cm [ca. 20"]; Tsehaye et al., 2013). Thus, for an invasivory campaign to have an impact on carp populations, fishers must achieve incredibly high harvest levels to reduce population size or there must be incentives for fishers to take smaller individuals. Likewise, the survival of adult wild boar (*Sus scrofa*) seems to have little impact on boar population growth rates, and the survival of piglets is the only life stage that may be likely to reduce boar population growth rates (Mellish et al., 2014). Thus, invasivore boar hunters would have more impact hunting piglets than the largest "trophy" animal, but this may be harder or less appealing to most hunters and pork consumers.

For many invasive plants, foragers may prefer to eat specific parts of the invasive plant. If foragers only remove portions of the plant, there may be little impact on the population. Garlic mustard, the wild forest herb, is a favorite early spring plant for foragers who like the spicy mustard flavor of the plant's leaves, stems, seeds, and roots. Garlic mustard is a biennial plant, which means that individuals live for two years before producing flowers and fruit. Some foragers prefer to harvest first-year plants when they are a small bunch of rosette leaves (Besemer, 2021), while others prefer the young spring leaves of second-year plants that are tender and less bitter (Rose, 2020). Other foragers like to snap off the tops of second-year plants before the plants produce flowers (Wesserle, 2017), and some like the spicy pop of the mustard's tiny black seeds (Besemer, 2021). Importantly, biologists estimate that managers need to harvest 85% of second-year adult plants or over 95% of first-year juvenile rosettes (Davis et al., 2006; Pardini et al., 2009; Evans et al., 2012). Harvest of individual leaves, or even snapping off just a plant's stalk, will likely not impact the plant's ability to produce seeds that year, and in some instances perpetuate the spread of seed to new locations. In fact, some foragers recognize that garlic mustard foraging is unlikely to impact future garlic mustard generations. One forager writes: "Garlic mustard is an invasive European species that has naturalized on four continents. You can harvest it freely without worrying about sustainability issues. You won't make a dent in this plant's population by eating it all year" (Meredith, 2020). Another, more avid, but vitriolic forager writes: "Some of my peers may say things like 'eat the invasives' . . . that's all well and good, and I appreciate the sentiment, but honestly, while plucking a few garlic mustard leaves here and there . . . might make you look like you're helping . . . you're more than likely not going to make an actual difference in the polyculture of your local woods" (Bergo, 2020).

For edible invasive species like garlic mustard and carp, the predominant downside in mismatched management and invasivory life stages is that using invasivory as a management tool may require more overall effort than other management campaigns that don't need to consider what life stages are the best for human consumption. Most concerning to invasivores should be instances when removing the tastiest individuals from a population leads to more, not fewer, of the invasive species. The biological term for a scenario where removal of some individuals leads to increased population growth rates is "overcompensation" (Schröder et al., 2014; Grosholz et al., 2021). This may be very common when an invasivory campaign focuses on harvesting the largest adults. Large adults can sometimes have a disproportionate impact on population growth rates, relative to smaller or younger individuals. Removal of large adults leads to overcompensation by stimulating the survival or growth of younger or smaller individuals. French-inspired invasivores searching for large American bullfrogs (*Rana catesbiana*) to batter and fry (Keller, 2013) may inadvertently increase the survival rates of bullfrog tadpoles because large male bullfrogs frequently cannibalize their tadpole offspring (Govindarajulu et al., 2005; Turner, 2016). Similarly, invasivores delighted by Mediterranean-inspired soft-shelled green crab (*Carcinus maenas*) or *masinette* crab caviar (https://www.greencrab.org/) may find that their harvest efforts could increase crab populations 30-fold if adult-juvenile crab cannibalism is prevalent within the population they are harvesting (Grosholz et al., 2021). Cannibalism aside, large adults can influence juvenile survival rates through other means. Larger adults can be aggressive and territorial and can physically exclude younger and smaller individuals from a habitat. Larger adults also are likely better at acquiring limited resources in a habitat, and thus can exclude younger individuals from these resources without direct physical interactions. For this reason, many population demographic models find that removal of larger, dominant individuals allows younger and smaller individuals to either grow faster, survive in greater number, or immigrate into prime habitat once the dominant individual is gone. This effect has been found for edible invasive crustaceans like green crabs (*Carcinus maenas*; Kanary et al., 2014; Gharouni et al., 2015) and signal crayfish (*Pacifastacus leniusculus*; Olsson et al., 2010; Moorhouse & Macdonald, 2011; Bohman et al., 2016), edible invasive fish like lionfish (*Pterois volitans* and *P. miles*; Barbour et al., 2011; Morris et al., 2011; Johnston & Purkis, 2015; Smith et al., 2017) and common carp (*Cyprinus carpio*; Weber et al., 2016), and the invasive mammal wild boar (*Sus scrofa*; Hanson et al., 2009; Klinger et al., 2011). For many edible invaders with a population-propensity for compensatory growth or overcompensation, invasivores who selectively harvest only bigger animals may find that each year they are easily able to fill their baskets just as the year before. For invasivores who are also hoping to do a good environmental deed through their diet, this would be a truly counterproductive outcome.

Population growth models can help anyone interested in managing invasive populations select the appropriate life stage for harvest. For managers with the sole intention of reducing invasive populations, they can adjust management plans and management tools to target the life stage with the highest likelihood of reducing the population size. For invasivores—who are interested in balancing enticing meals against reducing invasive populations—simply switching harvest strategies may not work. For all invasivores, harvesting and hunting smaller individuals can be harder and more time-consuming than focusing on the largest individuals. For commercial invasivores who would likely get paid per pound of harvest, it makes the most economic sense to bring in the largest individuals only. Additionally, in some instances, the most effective life stages for management—like bullfrog tadpoles or larval green crabs—are not currently considered palatable to most Americans. Overcoming American neophobia to new foods like these, or the economic inefficiency of harvesting smaller or less desirable individuals, could reduce excitement about eating invaders for population control. The fact that desirable and economically efficient harvests will often conflict with what is necessary for population reduction is a true catch-22.

Although the most popular invasivory narrative describes eating invaders eventually leading to population decline, we have provided a handful of examples where the biology of the invasive organism leads to unexpected, often counterproductive, outcomes. As we have already noted, the management of biological invasions through means other than invasivory is rife with examples of unintended consequences. Human release of biological control organisms has led to a myriad of negative ecological impacts (Simberloff, 2012). Likewise, removing single invasive species can sometimes lead to the unexpected release and expansion of other non-native species (Zavaleta et al., 2001; Ballari et al., 2016; Pearson et al., 2016). There is a rich literature on best practices in invasive species management that builds on past successes and learns from past failures (Lodge et al., 2006; Ricciardi et al., 2017), which any management program—invasivory or otherwise—should incorporate.

We argue that the examples of how some edible species' biology can lead to unintended consequences should elicit more caution from journalists, chefs, and biologists. Making widespread claims—without sound biological evidence supporting them—about the potential for invasivory campaigns to control invasive species could promote investment into recreational or commercial harvests that may lead to more, not fewer, invasive organisms. This result could come just from the biology of the organism. Later in the chapter, we discuss how perverse incentives, specifically arising from commercial markets, could also lead to spread or growth of invader populations. Unfortunately, we lack long-term demographic information for many edible invaders; perhaps the invasive species we discuss here are outliers, or perhaps they are harbingers of most invasivory-management campaigns. For the many species where we lack reliable population information, we encourage local foragers to use their fine-tuned observational skills to monitor how populations of invaders respond to harvest. Are populations getting smaller or bigger each year after harvest? Is there a visible impact of harvesting efforts? We also encourage biologists and environmentalists who promote invasivory as a "sustainable" way of eating to speak honestly to journalists, chefs, or budding invasivores when we have the biological evidence that the strategy will work, *and* when we do not (Pasko & Goldberg, 2014).

Is Invasivory Legal?

Although legal considerations are rarely noted in popular news articles about invasivory, navigating federal and state laws on the legality of harvesting, hunting, possessing, transporting, marketing, and selling wild-caught food is complicated. Broadly, at least in the United States, there are two bodies of law that may constrain invasivory activities: food safety laws and environmental laws (Galperin & Kuebbing, 2013). Food safety laws predominantly constrain commercialized invasivory activities because they are designed to protect consumers from eating unsafe or "adulterated" food products. However, environmental laws—many that are meant to prevent the further spread and impact of ecologically harmful invasive species—can apply to recreational and commercial invasivory activities. We analyze here the case of United States legal restrictions, but similar food safety and environmental laws are likely to exist in many other countries.

Food Safety Laws in the United States

Food safety laws impose food processing and inspection requirements to protect consumers from eating contaminated or unsafe food products. The primary regulators of food products in the United States are the U.S. Department of Agriculture (USDA)—which largely oversees meat and poultry production—and the U.S. Food and Drug Administration (FDA)—which oversees other food products. The USDA imposes stringent requirements on the sale of wild animal meat to insure it is "unadulterated" before human consumption. Thus, inspectors must often pre-inspect animals before slaughter and the slaughter must occur in government-approved facilities. This process, which certainly applies, for example, to invasive wild boar, will also likely limit other large mammals, like European rabbits (*Oryctolagus cuniculus*) and nutria (*Myocastor coypus*), or birds like wild turkeys (*Meleagris gallopavo*), that are popular invasivore ingredients.

Federal food safety restrictions have a history of preventing commercial invasivory markets. The New Orleans Chef Philippe Parola was an early advocate for invasivory in haute cuisine (Parola, 2022). Before the term "invasivory" was born, Chef Parola partnered with the Louisiana Department of Wildlife and Fisheries (LDWF) in the mid-1990s to demonstrate how restaurateurs and home cooks could serve nutria as a low-fat protein in Cajun cuisine (Ewbank, 2018; Conant, 2020; Parola, 2022). Biologists with LDWF were eager to engage the public in harvesting and eating nutria; aerial surveys of coastal marshlands throughout Louisiana estimated that the large rodents damaged over 102,585 acres of wetlands between 1998 and 2002 and were reducing crop yields in rice and sugarcane fields (Kinler et al., 1998; LDWF, 2002, 2022; Sasser et al., 2018). As with current day invasivore movements, Chef Parola's charisma and enthusiasm garnered widespread media attention (Parola, 2022). However, federal law stymied the Chef's attempts to build a nutria market because the law did not permit the sale or serving of nutria that hunters caught and killed in the wild (Barth, 2014; Ewbank, 2018).

While the law may take invasive mammals and birds off the table for commercial invasivory markets, food safety laws pose a lower barrier for invasive fish, shellfish, and plants. Fish harvesting is regulated by the FDA, and while FDA rules for seafood are substantial (FDA, 2022), the FDA does not require pre-mortem inspections of harvested fish (Fish and Fishery Products, 21 CFR § 123). The fishing industry also already has the legal infrastructure to navigate other food safety complexities—like preparing a detailed Hazard Analysis Critical Control Point Plan that ensures the safe handling and processing of harvested fish—that may make it easier to create a new market for invasive fish. For these reasons, it may be unsurprising that two of the most advanced invasivory markets—Asian carp in the Mississippi (Illinois Department of Natural Resources, 2018) and lionfish in the southeastern U.S. and Caribbean (Florida Fish and Wildlife Conservation Commission,

2022)—have successfully navigated food safety requirements. It is important to note that even while these are the largest invasivory markets, whether they are leading to declines in carp or lionfish populations is still very much an outstanding question (Andradi-Brown, 2019).

Food safety laws—designed to protect consumers from unsafe or unadulterated food products—can impact invasivory movements designed to stimulate economic opportunities for wild-food harvesters. For many edible invasive mammals and birds, USDA and FDA regulations on pre-harvest inspections and processing facilities may completely stall invasive food markets, as seen with historical attempts to grow a Louisiana wild-harvested nutria market. For edible invasive fish and plants, food safety requirements may raise the costs and infrastructure required for entrepreneurs engaging in an invasive food market, but not shut down sale of invasive food altogether.

Environmental Protection Laws

Both federal and state lawmakers have enacted laws specifically designed to reduce the ecological impact of invasive species. The purpose of these laws is to restrict the possession, transport, or sale of designated invasive species to reduce the spread and introduction of invasive species into uninvaded regions (Lacey Act, 16 U.S.C. §§ 3371-3378; Environmental Law Institute, 2002). Unlike food safety laws that may impose additional requirements on commercial markets for invasives or do not apply to recreational invasivores, environmental laws may totally block both commercial and recreational invasivore activities.

There are two primary federal laws governing possession and transportation of invasive species. The Lacey Act of 1909 (16 U.S.C. §§ 3371-3378) prohibits the transportation, sale, purchase, or acquisition of any animals the U.S. Fish and Wildlife Service (FWS) deems "injurious" to natural habitats. Similarly, The Plant Protection Act (7 U.S.C. §§ 7701-7786) prohibits the importation or transportation of plants the U.S. Department of Agriculture's Animal and Plant Health Inspection Service (APHIS) deems "noxious weeds." Some tasty invaders can be found on both these lists of regulated species: FWS currently lists the European rabbit (*Oryctolagus cuniculus*) and Atlantic salmon (*Salmo salar*), which are invasive in the northwestern U.S., as prohibited species. The law does not allow anybody, including invasivores, to transport, possess, or sell these species as live or dead animals. FWS also bans the sale and movement of living snakehead fish (*Channa argus*) and four species of Asian carp (*Cyprinus carpio, Hypophthalmichthys molitrix, H. nobilis,* and *Mylopharyngodon piceus*) or viable eggs (including viable eggs in dead, but fertile, female fish). APHIS also has a few edible "noxious weeds" on their banned plant list, including prickly pear (*Opuntia ficus-indica*) and Himalayan blackberry (*Rubus armeniacus*). Because both laws prevent the movement of listed species, they hamper commercial and recreational invasivore activities.

State environmental laws pose a more complex challenge for commercial and recreational invasivory advocates (Galperin & Kuebbing, 2013). Laws in nearly all 50 states establish rules about possession, transportation, sale, importation, exportation, and marketing of certain invasive species (Grove & Moltz, 2019). These laws, and their specifics, differ from state to state, creating a matrix of restrictions across the United States. For any invader of interest, recreational invasivores would need to locate their home state's invasive species lists to figure out which species they are allowed to harvest, possess, and transport, and which species the laws bar them from moving around. Commercial invasivores wishing to grow a market for a species that goes beyond state borders would need to map out which states would allow for the sale of invasive species for consumption and which states would ban such markets.

Some popular edible invasive species are also regulated in many states. The fruit-producing woody shrub autumn olive (*Elaeagnus umbellata*) produces copious bright red berries in the fall, rich in the nutrient lycopene, that invasivores make into jams and other tasty desserts (Stauffer, 2017; Maniez, 2020; Noyes, 2020). However, laws prohibit commercial and recreational invasivores from harvesting or moving autumn olive fruit within the states of Connecticut (Conn. Gen. Stat. § 22a-381d), Indiana (312 IAC 18-3-25), Maine (CMR 01-001-273), Massachusetts (Massachusetts Prohibited Plant List, 2022), New York (6 NYCRR § 575.3), Ohio (OAC Ann. 901:5-30-01), and West Virginia (W. Va. CSR § 61-14A-5). Similarly, the large knotweed shrubs (classified as *Polygonum* spp., *Reynoutria* spp., or *Fallopia* spp.) produce edible, asparagus-like red shoots that are popular as rhubarb substitutes in pies and desserts (Rose, 2016) or pickles (Preston, 2016). Knotweed harvest and transport is banned in Alabama (Ala. AAC § 80-10-14), California (3 CCR § 4500), Colorado (8 CCR § 1206-2), Connecticut (Conn. Gen. Stat. Ann. § 22a-381d), Idaho (IDAPA 02.06.09.220), Illinois (525 ILCS 10/3), Indiana (312 IAC 18-3-25), Maine (CMR 01-001-273), Massachusetts (974 Mass. Code Regs. 3.07), Montana (MONT. ADMIN. R. 4.5.207), Nebraska (Nebraska Admin. Code Title 25, Ch. 10), New Hampshire (Agr 3802.01 NH Prohibited Invasive Species), New York (6 NYCRR § 575.4), Ohio (OAC Ann. 901:5-37-01), Oregon (OAR 603-052-1200), Utah (U.A.C. R68-9-2),

Vermont (CVR 20-031-021), Washington (WAC § 16-750-011), West Virginia (W. Va. CSR § 61-14A-5), and Wisconsin (Wis. Adm. Code NR 40.04).

It is obvious that harvesting, eating, or selling some edible invaders is not as straightforward as just finding populations and picking (and selling). For any given invasive species, there may be a complex maze of federal food safety and federal and state environmental protection laws that limit the collection and harvest of these species. Invasivores should also pause if they happen to find themselves foraging in states without legal restrictions on certain invasive species. Species like knotweed and autumn olive are heavily regulated in some, but not all states. The states that do regulate the movement and spread of these plants do so because they recognize that moving viable plant parts, such as berries full of autumn olive seeds or shoots of invasive knotweeds that can re-root if dropped on the ground, is a way for these invasive plants to move to new habitats. States that lack these restrictions are more likely to do so because they lack strong environmental laws to prevent the spread of invasive species, not because the risks of spreading these species through harvesting activities are any lower.

To promote invasivory it might make sense to change laws to allow more wild harvest, consumption, and transportation of non-native, invasive species. But is deregulation the solution to biological invasions? It seems unlikely. First, even a deregulated invasivory market faces the biological hurdles that we discussed earlier in this chapter. Second, food safety and environmental regulations serve important purposes. While there is room for debate about the extent and nuance of these laws, the goal is to reduce risks to human health and to limit the spread of invasive species. Third, and finally, if successful, deregulating invasivory *might* allow invasivores to grow powerful markets, but as we discuss in the following section, powerful markets may not have the consequences invasivores expect.

What Happens If We Love It So Much, We Don't Want to Love It to Death?

There are two key aspects of invasivory campaigns that may prevent "eating invader" movements from effectively managing invasive species: integration into local culture and economic profitability (Nuñez et al., 2012, 2018). If a non-native, invasive species is incorporated into local cultures or cuisine, it then becomes a valued commodity and part of the local flora or fauna. Depending on how valued the non-native invasive becomes, recreational or commercial invasivores, or others in the local community, may actively avoid efforts to reduce or eradicate local populations of the beloved species. Additionally, when non-native, invasive species become culturally or economically valuable there are more likely to be debates among biologists, conservationists, and policymakers about the relative benefits versus harms of those species, which may reduce overall enthusiasm for any kind of management of the invasive species (Milanović et al., 2020; Vimercati et al., 2022). Similarly, commercial invasivores who wish to stimulate economic demand for an invasive species to encourage increased harvest and use of a species must create a profitable invasivore market, but that would incentivize people to maintain invasive populations, not reduce them. In short, a successful invasivory campaign would reduce or eradicate invasive populations, but use of markets and appeal to local culture may have competing goals and measures of success.

Cultural Integration of Edible Invaders

The cultural integration of some non-native, invasive species can be so deep that many people may not even realize that a beloved species is also an ecologically harmful invader (Nuñez & Simberloff, 2005; Nuñez et al., 2018). Wild horses are a staple cultural icon of the western United States (Edwards et al., 2012) and equid ancestors of today's wild horses (*Equus caballus*) once roamed North America (Librado & Orlando, 2021). Somewhere around 10,000 years ago, North American horses disappeared from the fossil record (Haile et al., 2009). For most of the Holocene, North American flora and fauna evolved without horses. When Spanish Conquistadors arrived in North America in the 15th century they introduced a domesticated descendent of North America's equine lineages (Librado & Orlando, 2021). These horses are beloved by many and protected by the federal Wild Free-Roaming Horses and Burros Act (Public Law 92-195), but also have large ecological impact on vegetation and wetlands across the American southwest (Norris, 2018; Scasta et al., 2018). There is even contentious debate on how to classify horses in North America, which may also be driven by the strong emotional connections between humans and horses. While some people contend that *E. caballus*'s equine ancestry lends the modern species "native" status in need of protection, other people wholeheartedly disagree, considering modern horses "invasive" and causing significant ecological and economic damage across much of the western U.S. (Klein, 2014; Masters, 2017). The cultural adoration of wild horses leads to conflict between land managers wishing to reduce population sizes of wild horses and people advocating for their conservation (Scasta et al., 2018; American Wild Horse Campaign [https://americanwildhorsecampaign.org]; Return to Freedom [https://returntofreedom.org/about/our-mission/]). Interestingly, while American chefs and consumers do not buy wild horse meat for human consumption, horse meat markets for human consumption exist in Europe, Japan, and Russia. One of the most heated controversies in the wild horse protection versus management debate is that many of the wild horses collected by federal land managers are shipped to international horse meat markets for sale for human consumption (Bloch, 2019).

In other instances, more palatable invasive species are integrated into American culture and cuisine, but with wider recognition of their ecological impacts. Wild boar (*Sus scrofa*) are a damaging non-native invader that European colonists introduced. Wild boar can extirpate native plant populations and change forest soils through their extensive burrowing and wallowing activities (Barrios-Garcia & Ballari, 2012). In 1982, only 17 states, predominantly in the southern U.S., reported feral populations of wild boar; in 2021, the count had nearly doubled to 31 states including areas of Vermont, New Hampshire, Michigan, and North Dakota (USDA APHIS, 2021). Population models of boar predict that in 2016 the U.S. boar population was ~2.4 million pigs, but based on habitat availability and changing climates, this number could increase twelve-fold to 21.4 million animals in the future (Lewis et al., 2019). The U.S. Department of Agriculture estimates that feral swine cause approximately $1.5 billion in damage and control costs each year, with at least $800 million of those damages stemming from damage to agricultural crops and lost yields (USDA APHIS, 2020; USDA APHIS, 2022).

The widespread impacts of wild boar have turned them into infamous cultural icons for America's recreational hunting community. Boar hunters host wildly popular social media accounts that share their adventures and successes tracking and shooting boar. Facebook hosts numerous popular wild boar hunting pages including Wild Boar Hunting (@boarhunterno1, https://www.facebook.com/boarhunterno1/) with over 109,000 followers. YouTube has a large selection of hunting videos including "Hunting HOGZILLAS in East Tennessee," which has been viewed over 1.2 million times since its posting in October of 2021 (https://www.youtube.com/watch?v=288sJXakEc8), and "Solo California Wild Pig Hunt," which has been viewed nearly 90,000 times (https://www.youtube.com/watch?v=yt_KejME44U). Wild boar are also the star antagonists in the American reality television show *American Hoggers*, which follows a "legendary hog hunter Jerry Campbell, and his level-headed son, Robert, 28 and firecracker daughter, Krystal, 23" who track and hunt wild boar that are "terrorizing helpless landowners" (IMDB, 2021). And wild boar have even made their way into episodes of celebrity chef cooking shows. British chef Gordon Ramsey visited Georgia to hunt and prepare wild boar on a U.S. military base, sharing his adventures in an episode of his popular show *The F Word* (https://www.youtube.com/watch?v=6p5x0nxtqVs).

Boars, and more importantly, boar hunting, are clearly popular in the United States. For this reason, advocates of invasivory argue that increased hunting and consumption efforts could control wild boar populations. Hunter, chef, and invasivore Jackson Landers writes: "The [wild boar] problem isn't going to be solved without bringing some real weapons to bear. We need new hunters to step up to the plate and start hunting the invasive pigs in their own areas. The existing hunting subculture probably can't do this on its own" (Landers, 2012a). A slaughterhouse in Springfield, Louisiana, is now selling wild boar meat—caught and brought live to the slaughterhouse for inspection (per USDA restrictions, see above)—to New Orleans restaurants eager to serve boar bacon at brunch or boar charcuterie and meatballs for dinner (McConnaughey, 2017).

Unfortunately, Americans' love of boar hunting is so strong that instead of curtailing boar populations, hunting is propelling the boar's expansion (Tabak, 2017; Lewis et al., 2019). Hunters interested in tracking animals on their own property will capture live boar and relocate the animals to new areas without boar populations. It is most likely that the scattershot spread of boar within the U.S. is a result of these human-assisted translocations (Lewis at al., 2019). This penchant for translocation is the justification for many state legislatures creating policies and laws banning or severely restricting the movement of living wild boar (Johnson v. DNR, 2015). The story arc of wild boar in the United States serves as an important example that loving invasive species can very easily lead to the proliferation of the species, and that commercializing the wild harvest of invasive species may overwhelm goals of reducing or managing invasive populations.

Although invasive wild boar might be the prime (meat) example of how cultural incorporation of an invasive species can lead to its spread, it's not the only edible invasive that is woven into the iconography of the United States. Kudzu (*Pueraria montana* var. *lobata*) is a rapidly growing, lush green vine that has become a regional icon for (and on) roadsides in the southeastern United States (Alderman, 2015; Solomon, 2021). Kudzu symbolism can be found in the title of Florida State University's Undergraduate Literary Magazine (*The Kudzu Review*, https://kudzureviewfsu.com), the lyrics of country ballads sung by Johnny Cash ("The L&N Don't Stop Here Anymore," https://www.youtube.com/watch?v=0jF8QgyW3VA) and Florida Georgia Line ("Countryside," https://www.youtube.com/watch?v=X2ga2CvQIgo), and the foliar foe encasing and damaging the skeletal remains of a victim in the hit crime-drama television show *Bones* ("The Carrot in the Kudzu," Season 9 Episode 18, https://www.imdb.com/title/tt3547454/). Today, architects, artists, and chefs are eager to incorporate this cultural icon into their work (Alderman, 2015; Horn-Muller, 2021). On the opposite coast, Himalayan blackberry (*Rubus armeniacus*) has taken firm root in Seattle, Washington's cultural landscape. The prickly, fruit-producing vine was featured heavily in American author Tom Robbins's novel *Still Life with Woodpecker* (Robbins, 1980) as the encapsulating prickly vine surrounding the Seattle mansion of a royal family. Seattle meteorologist Shannon O'Donnell for KOMO News shared with her 13.7 thousand followers on Twitter that "Nothing says 'late summer' near Seattle like the explosion of non-native Himalayan blackberries.... Considered a noxious weed, they may be thorny & invasive, but sure are juicy & delicious!" (https://twitter.com/ShannonODKOMO/status/1428172222624006148). Whether beloved or notorious, edible invasives have already made inroads into American culture. The question is whether Americans want these icons to be enduring or endangered.

Economic Profitability of Edible Invaders

Many edible non-native species that are invasive today were originally introduced intentionally into their non-native range because humans wanted to cultivate the species for food. The first large wave of introduced species to North America came by boat with British colonists in the 17th and 18th centuries. It is likely that colonists introduced many edible European plant species including broadleaf plantain (*Plantago major*), Queen Anne's lace (*Daucus carota*), curly dock (*Rumex crispus*), dandelion (*Taraxacum officinale*), and sweet fennel (*Foeniculum vulgare*) into gardens as medicinal or food plants (De Schweinitz, 1832; Mack, 2003). Likewise, early European colonists intentionally released wild boar (*Sus scrofa*) and European rabbits (*Oryctolagus cuniculus*) to North America wherever their boats landed to stock the Americas with familiar game species (Hodge & Lewis, 1907; Hanson & Karstad, 1959; Flux & Fullagar, 1992; Flux, 1994).

Later, as European colonists and their descendants ate their way across the continent, they saw tasty native species going locally extinct owing to overharvesting, introduction of novel diseases, loss of habitat, or a combination of factors. As edible native species declined, local communities replaced them with edible non-natives. The California Gold Rush of 1849 enticed large numbers of people to emigrate to San Francisco and surrounding counties. The newly arrived immigrants, many of whom left France for California, were fond of frog legs and found the native Californian red-legged frog (*Rana aurora draytonii*) to be quite tasty. Wildlife harvesting records show heavy commercial demand for red-legged frogs throughout northern California up until 1900, when it appears that hungry Californians had depleted the native frog population and businesses were importing frogs from out-of-state to meet consumer demands (Jennings & Hayes, 1985). As native red-legged frog populations declined, entrepreneurial frog farmers began digging ponds and stocking their waters with common bullfrogs (*Rana catesbeiana*) imported from eastern North America (Storer, 1922; Jenning & Hayes, 1985). The non-native bullfrogs did well, eventually escaping from frog farms and invading aquatic habitats across the western United States.

A similar story exists for wild turkeys (*Meleagris gallopavo*), which hunters and habitat loss extirpated from an estimated 88% of their native range in North America (Mosby, 1949). Aggressive reintroduction programs in the mid-20th century were successful in reestablishing native turkey populations, and by 1974 turkey populations were robust and had recovered across ~25% of their native range (Mosby, 1975). Today, wild turkeys are not only populous throughout their original native range, but also into new habitats well outside their original range. The overzealous stocking efforts of the bird in California, a region with no historical populations of this species, have led to concerns about the species' ecological impact on native Californian species (California Department of Fish and Game, 2004; Morrison et al., 2016).

There are many other examples of humans moving species outside their historical native North American range to meet growing market demands. Invasive Atlantic salmon (*Salmo salar*) can be found in freshwater and marine habitats throughout most of the western United States. Some of these Atlantic salmon introductions were accidental, when farmed fish escaped their aquatic pens in the Pacific Ocean (Naylor et al., 2001). Other Atlantic salmon introductions are intentional, as stocked fish for sport fisheries (Bahis, 1992). Similarly, blue catfish (*Ictalurus furcatus*) have been stocked for sport and food fisheries well beyond their native range in the southeastern United States. Despite noted impacts on native fish species, many state fish and wildlife agencies manage blue catfish populations to sustain, not diminish, their populations (Graham, 1999; Fuller & Neilson, 2014).

The historical demand for invasive species as food resources has multiple implications for future invasivory markets. Encouragingly, past interest in eating invasives suggests that invasivores could recreate interest in consuming large numbers of these species. Ominously, these examples serve as warnings for how economic incentives for harvesting and selling invasives can do more harm than good by promoting the spread of edible invasive species as eager entrepreneurs seek entry into profitable marketplaces.

Today, invasivores seek to build markets for edible invasives to promote social benefits and increase environmental protection. Many of the proposed candidates for invasive species markets are prolific reproducers, already widespread across the United States, and with well-documented ecological impacts. Current disinterest from the public and policymakers and limited funding for invasive species management have stymied efforts to curtail invasive populations and reduce their impacts.

The rub with the creation of economic markets to control invasive species, therefore, is that these markets must be large and lucrative to realistically reduce most invasive species populations (Harris et al., 2023). As discussed earlier in this chapter, harvest rates for many edible invasive species must be quite high—well over 50% of the population size each year and sometimes even over 90%—to cause a decline in population growth rates. Yet, large and lucrative markets are unwieldly. Once a market exists, there is no economic or social tool that a well-intentioned invasivore market-creator can employ that would dictate or control the actions of future invasivores motivated more by profit than social good. As an invasivory market grows larger and more lucrative, there will likely be more individuals wishing to invest in the market. Invasivores must ask themselves: when the demand for the species grows, what will stop suppliers from taking actions that will increase their profits and spread a species, instead of causing its decline? What will keep a commercial invasivore from moving the invasive species to new locations that are easier to access or closer to other infrastructure that will increase their profits, as we have already seen with wild boar? What will stop innovative suppliers from figuring out how to farm an invasive species, like American bullfrogs or Atlantic salmon, which would surely be easier to harvest and sell than tromping through the forests and fields to harvest wild invaders? How do you ensure that commercial invasivores collecting wild berries from invasive autumn olive (*Elaeagnus umbellata*), reproductive young shoots of knotweed (*Polygonum* spp.), or seeds of garlic mustard (*Alliaria petiolata*) will safely transport their harvests in ways that don't accidentally spread the species on its way from field to market? These are critical questions that any advocate of commercial invasivory programs should address.

Conclusion

Everybody loves a good "win-win" narrative. Unfortunately, "win-wins" are likely elusive because environmental problems arise from many interacting social, cultural, and economic complexities making it unlikely that a single action can lead to benefits for all stakeholders (Hegwood et al., 2022). Invasivory has been broadly pitched as a "win-win" for creative chefs, avid foragers and hunters, and invasive species managers alike, without much regard to biological, legal, and socioeconomic complexities that may stall or counteract the goal of reducing invasive species populations. Here, we review those complexities and provide many examples where it appears unlikely that invasivory can be both a "win" for human diets and a "win" for population reduction.

Even if many "eating invader" campaigns will fall short of meeting the dual objectives of harvesting invasive species for food and controlling their populations, "eating invaders" is not inherently a bad thing. Recreational invasivory focused on foraging, harvesting, and hunting from local populations can have many benefits that are a "win" for the forager's table and home-cooked meals. Invasivore chefs looking for an interesting new ingredient may also have "winning" menus that delight and educate diners about the impacts of invasive species, such as that of Connecticut chef Bun Lai who was honored as a "White House Champion of Change for Sustainable Seafood," in part because of his use of invaders like green crabs and lionfish on his menu (White House, 2016). Yet, population demographic models demonstrate the biological realities that most invasive species management campaigns face: reductions in population size come from incredibly high harvest levels of specific individuals of certain ages or sizes. Recreational invasivores may be able to reduce the population size of small, satellite populations of invaders in their backyard or local natural area, but recreational invasivory is unlikely to achieve the high harvest rates necessary to control widespread invaders across county and state borders. Additionally, recreational invasivores may find that their favorite edible plant parts or large adult animals may not be the most appropriate individuals to harvest for population reduction. There is limited evidence that recreational invasivory will be a "win" for managing invaders.

Far from being a win-win, the idea of commercial invasivory poses a genuine environmental threat and advocates must approach it with skepticism. Commercialized invasivory—or invasivores who are driven to harvest as many individuals as possible because of economic incentives—may be able to harvest and remove enough invasive individuals to drive down population sizes by drumming up public demand. But the benefit of widespread harvest comes with challenges. With commercialized invasivory and high public demand comes the chance that the public will grow to love an invader, and thus be reluctant to see its population diminish. This says nothing of those with commercial investments in the invasivore market who will be faced with the choice of shrinking invasive populations or growing their own wealth. With commercialized invasivory come more legal challenges, like the shutdown of a nutria meat market, which will require more financial investment in well-planned campaigns that adhere to federal and state laws regulating food safety and environmental protection. With commercialized invasivory and high public demand for a consumable product come more participants in a marketplace that may no longer be driven by the social good to reduce the invasive population. We show examples of profiteering in an economic marketplace that can lead to the spread of invasive species like boar, bullfrogs, and blue catfish, by individuals wishing to make money off the invaders, not to "*eat 'em to beat 'em.*"

There are examples, however, where economic incentives have been successful *at engaging public participation in invasive species control.* Tenacious biologists at the Louisiana Department of Wildlife and Fisheries (LDWF) explored multiple avenues for controlling widespread and populous nutria in Louisiana's swamps. In addition to engaging with Chef Philippe Parola to stimulate a nutria meat market, they also tried to grow an international market for nutria fur. This endeavor also failed because nutria pelt prices were never high enough to stimulate enough nutria trappers to engage in the market (LDWF, 2002). After two unsuccessful attempts at growing commercial product markets for nutria, the State gave up on pursuing international markets.

LDWF instead turned to a bounty harvest-incentive program to get public help in controlling nutria population (LDWF, 2002). Today, hunters enrolled in the Coastwide Nutria Control Program get paid $6 per nutria tail, and in the winter 2020–2021 hunting season 284 hunters deposited 312,118 nutria tails to state wildlife biologists. Ninety-nine percent of the hunters reported to wildlife officials that they disposed of nutria carcasses through three approved measures: burying the carcass, placing the carcass in overhead vegetation, or sinking the carcass in water; fewer than 1% reported keeping whole carcasses for other uses, such as consumption (Manuel & Waller, 2021). Like LDWF, the Florida Fish and Wildlife Conservation Commission (FFWCC) has also turned toward bounty programs—in the form of public derbies or fishing tournaments that bring people together in a single day to harvest and catch lionfish from Floridian reefs. Florida wildlife officials can more carefully regulate derby harvesters by dictating how and where they fish and ensuring fishers have the proper training to engage safely in the harvesting. In 2019, FFWCC reported that 21 tournaments resulted in the verified capture of 25,000 lionfish compared to only 1,233 fish from commercial lionfish operations (Florida Fish and Wildlife Conservation Commission, 2022).

Unlike private commercial markets, a single entity (usually a governmental body) controls bounty programs and sets rules for the harvest. While bounty programs for invasive pests have also stimulated bad-actors seeking economic profit to farm and spread invasives (Pasko & Goldberg, 2014), there is a single authoritative body that can curtail bad-actors thus providing more control in these restricted marketplaces. Thus, incentive programs can stimulate public participation in invasive species control, but still be closely monitored by governmental agencies who have the final say on the sale price and the existence of these types of invasive markets and can closely monitor public participation (Pasko & Goldberg, 2014).

Individuals who want to reduce the negative impacts of invasive species are the champions of invasivory. They want to engage others in conversations about how we can solve environmental problems (Roman, 2004; Gorman, 2010; Landers, 2012b; Conant, 2020; Parola, 2022; Seaman et al., 2021). These are noble underpinnings to invasivory campaigns, and also the reason that invasivory champions must advocate for more nuanced narratives about the benefits and challenges of eating invaders. These nuanced narratives are just beginning to percolate in scientific (http://eattheinvaders.org/faq/), culinary (Bergo, 2020), and journalistic (Mishan, 2020) writing. We hope that trend continues.

BOX 1 List of Cookbooks and Websites with Invasive Species Menus

COOKBOOKS

They're Cooked: Recipes to Combat Invasive Species
by Corinne M. Duncan, Melanie Gisler, and Tamara S. Mullen; Institute for Applied Ecology, 2014, 117 pages

Species featured: Asian carp, bullfrog, crayfish, dandelion greens, garlic mustard, Himalayan blackberry, house sparrow, Japanese knotweed, kudzu, nutria, purple varnish clams, sheep sorrel, starling, wild boar, and wild turkey

https://appliedeco.org/product/theyre-cooked-recipes-to-combat-invasive-species/

The Green Crab Cookbook: An Invasive Species Meets a Culinary Solution
by Mary Parks and Thanh Thái; Green Crab R&D, 2019, 127 pages

Species featured: green crab

https://www.greencrab.org/cookbook

Eat the Problem
by Kirsha Kaechele; Mona, 2019, 544 pages

https://shop.mona.net.au/products/eat-the-problem

How to Eat Green Crab: A Cookbook of Invasive Species Harvested along the Maine Coast
by Oliver Curtis; Telling Room Press, 2017, 48 pages

Species featured: green crab

https://www.tellingroom.org/bookstore/how-to-eat-green-crab-a-cookbook-of-invasive-species-harvested-along-the-maine-coast

The Invasive Species Cookbook: Conservation through Gastronomy
by J. M. Franke; Bradford Street Press, 2007 (out of print), 111 pages

Green Fig and Lionfish: Sustainable Caribbean Cooking (A Gourmet Foodie Gift)
by Allen Susser; Books & Books Press, 2019, 224 pages

Species featured: lionfish

https://shop.booksandbooks.com/book/9781642501643

The Lionfish Cookbook: The Caribbean's New Delicacy
by Tricia Ferguson and Lad Akins; Reef Environmental Education Foundation, 2016 (out of print), 161 pages (first edition: 2010)

Species featured: lionfish

The Book of Kudzu: A Culinary and Healing Guide
by William Shurtleff and Akiko Aoyagi; CreateSpace Independent Publishing Platform, 2018, 104 pages (previous editions: Autumn Press, 1977; Avery Publishing Group Inc., 1985; Soyinfo Center, 1998)

Species featured: kudzu

WEBSITES

Copi: https://choosecopi.com
GreenCrab.org: https://www.greencrab.org
Eat the invaders: http://eattheinvaders.org
Wild Man Steve Brill: https://www.wildmanstevebrill.com

Literature Cited

Alderman, D. H. 2015. "When an exotic becomes native: Taming, naming, and kudzu as regional symbolic capital." *Southeastern Geographer* 55: 32–56.

Andradi-Brown, D. A. 2019. "Invasive lionfish (*Pterois volitans* and *P. miles*): Distribution, impact, and management." Pp. 931–941 *in* Y. Loya, K. Puglise & T. Bridge (editors), *Mesophotic Coral Ecosystems. Coral Reefs of the World*, Vol. 12. Springer Nature, Cham.

Bahis, P. 1992. "The status of fish populations and management of high mountain lakes in the western United States." *Northwest Science* 66: 183–193.

Ballari, S. A., S. E. Kucbbing & M. A. Nuñez. 2016. "Potential problems of removing one invasive species at a time: A meta-analysis of the interactions between invasive vertebrates and unexpected effects of removal programs." *PeerJ* 4: e2029.

Barbour, A. B., M. S. Allen, T. K. Frazer & K. D. Sherman. 2011. "Evaluating the potential efficacy of invasive lionfish (*Pterois volitans*) removals." *PLoS ONE* 6: e19666.

Barnes, M. A., A. M. Deines, R. M. Gentile & L. E. Grieneisen. 2014. "Adapting to invasions in a changing world: Invasive species as an economic resource." Pp. 326–344 *in* L. H. Ziska & J. S. Dukes (editors), *Invasive Species and Global Climate Change*. CABI, Wallingford.

Barrios-Garcia, M. N. & S. A. Ballari. 2012. "Impact of wild boar (*Sus scrofa*) in its introduced and native range: A review." *Biological Invasions* 14: 2283–2300.

Barth, B. 2014. "The joy of cooking invasive species." *Modern Farmer*, 27 March 2014. https://modernfarmer.com/2014/03/invasive-diet-takes/

Bergo, A. 2020. "Garlic mustard: A Dangerous Invasive Edible." *Forager | Chef*, May 2020. https://foragerchef.com/garlic-mustard/

Besemer, T. 2021. "Garlic mustard—The tastiest invasive species you can eat." *Rural Sprout*, 3 March 2021. https://www.ruralsprout.com/garlic-mustard/

Bloch, S. 2019. "How America's wild horses end up in slaughterhouses abroad." *The Counter*, 9 September 2019. https://thecounter.org/americas-growing-horse-slaughter-trade/

Bohman, P., L. Edsman, A. Sandström, P. Nyström, M. Stenberg, P. Hertonsson & J. Johansson. 2016. "Predicting harvest of non-native signal crayfish in lakes—A role for changing climate?" *Canadian Journal of Fishery and Aquatic Science* 73: 785–792.

California Department of Fish and Game. 2004. "Strategic plan for wild turkey management." State of California, The Resources Agency, Department of Fish and Game. https://nrm.dfg.ca.gov/FileHandler.ashx?DocumentID=83157

Conant, E. 2020. "One way to fight invasive species? Eat them." *National Geographic*, 23 December 2020. https://www.nationalgeographic.com/travel/article/eating-invasive-species-on-a-road-trip-across-the-southern-us

Davis, A. S., D. A. Landis, V. Nuzzo, B. Blossey, E. Gerber & H. L. Hinz. 2006. "Demographic models inform selection of biocontrol agents for garlic mustard (*Alliaria petiolata*)." *Ecological Applications* 16: 2399–2410.

de Lapparent Alvarez, A. 2022. "Feasting on New York City's greenery." *Columbia News Service*, 3 May 2022. https://columbianewsservice.com/2022/05/03/feasting-on-new-york-citys-greenery/

De Schweinitz, L. D. 1832. "Remarks on the plants of Europe which have become naturalized in a more or less degree, in the United States." *Annals of the Lyceum of Natural History of New York* 3: 148–155.

The Economist. 2022. "Copicat: Rebranding the Asian carp." *The Economist*, 9 July 2022, p. 24 (United States).

Edwards, P., K. A. E. Enenkel & E. Graham. 2012. *The Horse as a Cultural Icon*. Koninklijke Brill NV, Leiden.

Environmental Law Institute. 2002. "Halting the invasion: State tools for invasive species management." August 2002. https://www.eli.org/sites/default/files/eli-pubs/d12-06.pdf

Evans, J. A., A. S. Davis, S. Raghu, A. Ragavendran, D. A. Landis & D. W. Schemske. 2012. "The importance of space, time, and stochasticity to the demography and management of *Alliaria petiolata*." *Ecological Applications* 22: 1497–1511.

Ewbank, A. 2018. "The chef cooking up invasive species." *Atlas Obscura*, 13 March 2018. https://www.atlasobscura.com/articles/eating-asian-carp-invasive-species

Executive Order 13112. 1999. Invasive Species. Federal Register, Vol. 64, Number 25, 8 February 1999. https://www.federalregister.gov/documents/1999/02/08/99-3184/invasive-species

FDA. 2022. "Seafood guidance documents & regulatory Information." U.S. Food & Drug Administration. https://www.fda.gov/food/guidance-documents-regulatory-information-topic-food-and-dietary-supplements/seafood-guidance-documents-regulatory-information

Florida Fish and Wildlife Conservation Commission. 2022. "Lionfish." https://myfwc.com/fishing/saltwater/recreational/lionfish/

Flux, J. E. C. 1994. "World distributions." Pp. 7–17 *in* H. V. Thompson & C. M. King (editors), *The European Rabbit: The History and Biology of a Successful Colonizer*. Oxford University Press, Oxford.

Flux, J. E. C. & P. J. Fullagar. 1992. "World distribution of the rabbit *Oryctolagus cuniculus* on islands." *Mammal Review* 22: 151–205.

Fuller, P. & M. Neilson. 2014. "*Ictalurus furcatus*." USGS Nonindigenous Aquatic Species Database, Gainesville. http://nas.er.usgs.gov/queries/FactSheet.aspx?speciesID=740 [revision date: 3 May 2021]

Galperin, J. U. & S. E. Kuebbing. 2013. "Eating invaders: Managing biological invasions with fork and knife?" *Natural Resources & Environment* 28: 41–44.

Garvey, J. E., G. G. Sass, J. Trushenski, D. Glover, M. K. Brey, P. M. Charlebois, J. Levengood, et al. 2015. "Fishing down the bighead and silver carps: Reducing the risk of invasion to the Great Lakes. Final Report." Technical Report, Southern Illinois University, Carbondale Illinois. https://www.researchgate.net/publication/281465022_Fishing_Down_the_Bighead_and_Silver_Carps_Reducing_the_Risk_of_Invasion_to_the_Great_Lakes

Gharouni, A., M. A. Barbeau, A. Locke, L. Wang & J. Watmough. 2015. "Sensitivity of invasion speed to dispersal and demography: An application of spreading speed theory to the green crab invasion on the northwest Atlantic coast." *Marine Ecology Progress Series* 541: 135–150.

Gorman, J. 2010. "A diet for an invaded planet: Invasive species." *New York Times*, 31 December 2010. https://www.nytimes.com/2011/01/02/weekinreview/02gorman.html?_r=0

Govindarajulu, P., R. Altwegg & B. R. Anholt. 2005. "Matrix model investigation of invasive species control: Bullfrogs on Vancouver Island." *Ecological Applications* 15: 2161–2170.

Graham, K. 1999. "A review of the biology and management of blue catfish." Pp. 37–49 in E. R. Irwin, W. A. Hubert, C. F. Rabeni, H. L. Schramm Jr. & T. Coon (editors), *Catfish 2000: Proceedings of the International Ictalurid Symposium*, American Fisheries Society Symposium 24. American Fisheries Society, Bethesda.

Grosholz, E., G. Ashton, M. Bradley, C. Brown, L. Ceballos-Osuna, A. Chang, C. de Rivera, et al. 2021. "State-specific overcompensation, the hydra effect, and the failure to eradicate an invasive predator." *Proceedings of the National Academy of Sciences* 118: e2003955118.

Grove, S. & M. Moltz. 2019. "Legislative and regulatory efforts to control invasive species." The Center for Rural Pennsylvania. https://www.rural.pa.gov/getfile.cfm?file=Resources/PDFs/research-report/Invasive-Species-Report-2019.pdf&view=true

Haile, J., D. G. Froese, R. D. E. MacPhee, R. G. Roberts, L. J. Arnold, A. V. Reyes, M. Rasmussen, et al. 2009. "Ancient DNA reveals late survival of mammoth and horse in interior Alaska." *Proceedings of the National Academy of Sciences* 106: 22352–22357.

Hajek, A. E., B. P. Hurley, M. Kenis, J. R. Garnas, S. J. Bush, M. J. Wingfield, J. C. van Lenteren & M. J. W. Cock. 2016. "Exotic biological control agents: A solution or contribution to arthropod invasions?" *Biological Invasions* 18: 953–969.

Halliday, T. 1980. "The extinction of the passenger pigeon *Ectopistes migratorius* and its relevance to contemporary conservation." *Biological Conservation* 17: 157–162.

Hanson, L. B., M. S. Mitchell, J. B. Grand, D. B. Jolley, B. D. Sparklin & S. S. Ditchkoff. 2009. "Effect of experimental manipulation on survival and recruitment of feral pigs." *Wildlife Research* 36: 185–191.

Hanson, R. P. & L. Karstad. 1959. "Feral swine in the Southeastern United States." *The Journal of Wildlife Management* 23: 64–74.

Harris, H. E., W. F. Patterson III, R. N. M. Ahrens, M. S. Allen, D. D. Chagaris & S. L. Larkin. 2023. "The bioeconomic paradox of market-based invasive species harvest: A case study of the commercial lionfish fishery." *Biological Invasions* 25: 1595–1612. https://doi.org/10.1007/s10530-023-02998-5

Hegwood, M., R. E. Langendorf & M. G. Burgess. 2022. "Why win-wins are rare in complex environmental management." *Nature Sustainability* 5: 674–680.

Herzog, K. 2011. "Chef creates meals featuring invasive Asian carp." *Milwaukee Journal Sentinel*, 23 January 2011. https://www.mprnews.org/story/2011/01/23/dinner-asian-carp

Hodge, F. W. & T. H. Lewis. 1907. *Original Narratives of Early American History: Spanish Explorers in the Southern United States 1528–1543.* Charles Scribner's Sons, New York.

Hongoltz-Hetling, M. 2021. "Welcome to invasivorism, the boldest solution to ethical eating yet." *Popular Science*, 7 December 2021. https://www.popsci.com/environment/eating-invasive-species/

Horn-Muller, A. 2021. "The role of kudzu in architecture, cuisine, and culture." *Southerly Magazine*, 20 March 2021. https://southerlymag.org/2021/03/20/the-role-of-kudzu-in-architecture-cuisine-and-culture/

Illinois Department of Natural Resources. 2018. "Asian Carp Business Process Analysis Final Report and Action Plan." 11 January 2018. https://www.ifishillinois.org/programs/AsianCARPReport.pdf

IMDB. 2021. *American Hoggers.* https://www.imdb.com/title/tt1988897/plotsummary?ref_=tt_ov_pl

IPBES. 2019. "Summary for policymakers of the global assessment report on biodiversity and ecosystem services of the Intergovernmental Science-Policy Platform on Biodiversity and Ecosystem Services." S. Díaz, J. Settele, E. S. Brondízio, H. T. Ngo, M. Guèze, J. Agard, A. Arneth, et al. (editors). IPBES secretariat, Bonn.

Jennings, M. R. & M. P. Hayes. 1985. "Pre-1900 overharvest of California red-legged frogs (*Rana aurora draytonii*): The inducement for bullfrog (*Rana catesbeiana*) introduction." *Herpetologica* 41: 94–103.

Johnson v. DNR. 873 N.W.2d 842 (Mich.App. 2015). https://www.courts.michigan.gov/496df3/siteassets/case-documents/uploads/opinions/final/coa/20150602_c321337(54)_rptr_69o-321337-final.pdf

Johnston, M. W. & S. J. Purkis. 2015. "A coordinated and sustained international strategy is required to turn the tide on Atlantic lionfish invasions." *Marine Ecology Progress Series* 533: 219–235.

Kanary, L., J. Musgrave, R. C. Tyson, A. Locke & F. Lutscher. 2014. "Modelling the dynamics of invasion and control of competing green crab genotypes." *Theoretical Ecology* 7: 391–406.

Keller, S. J. 2013. "Gnawing away at invasives." *High Country News* 45: 3.

Kinler, N., G. Linscombe & S. Hartley. 1998. "A survey of nutria herbivory damage in coastal Louisiana in 1998." Nutria Harvest and Wetland Demonstration Project, Fur and Refuge Division, Louisiana Department of Wildlife and Fisheries. https://www.lacoast.gov/reports/project/3891531~1.pdf

Klein, K. 2014. "Opinion: Is America's wild horse an invasive species, or a reintroduced native?" *Los Angeles Times*, 3 July 2014. https://www.latimes.com/nation/la-ol-wild-horse-endangered-20140703-story.html

Kleiner, M. 2019. "Sushi's role." *Yale Daily News*, 1 February 2019. https://yaledailynews.com/blog/2019/02/01/sushis-role/

Klinger, R., J. Conti, J. K. Gibson, S. M. Ostoja & E. Aumack. 2011. "What does it take to eradicate a feral pig population?" Pp. 78–86 *in* C. R. Veitch, M. N. Clout & D. R. Towns (editors), *Island Invasives: Eradication and Management.* IUCN, Gland.

Landers, J. 2012a. "Want to help the environment? Go shoot a pig." *Slate Magazine*, 9 August 2012. https://slate.com/technology/2012/08/hunt-wild-pigs-for-the-environment-kill-and-eat-invasive-species.html

Landers, J. 2012b. *Eating Aliens: One Man's Adventures Hunting Invasive Animal Species*. Storey Publishing, North Adams, Massachusetts.

LDWF. 2002. "Nutria in Louisiana." Louisiana Department of Wildlife and Fisheries and Genesis Laboratories, Inc. https://nutria.com/wp-content/uploads/0232.brochurerev.pdf

LDWF. 2022. "Herbivory damage and harvest maps." https://nutria.com/nutria control-program/herbivory-damage-and-harvest-maps/

Lewis, J. S., J. L. Corn, J. J. Mayer, T. R. Jordan, M. L. Farnsworth, C. L. Burdett, K. C. VerCauteren, S. J. Sweeney & R. S. Miller. 2019. "Historical, current, and potential population size estimates of invasive wild pigs (*Sus scrofa*) in the United States." *Biological Invasions* 21: 2173–2384.

Librado, P. & L. Orlando. 2021. "Genomics and the evolutionary history of equids." *Annual Review of Animal Biosciences* 9: 81–101.

Linnekin, B. J. 2018. "Food law gone wild: The law of foraging." *Fordham Urban Law Journal* 45: 995–1050.

Liu, C., L. Comte & J. D. Olden. 2017. "Heads you win, tails you lose: Life-history traits predict invasion and extinction risk of the world's freshwater fishes." *Aquatic Conservation: Marine Freshwater Ecosystems* 27: 773–779.

Lodge, D. M., S. Williams, H. J. MacIsaac, K. R. Hayes, B. Leung, S. Reichard, R. N. Mack, et al. 2006. "Biological invasions: Recommendations for U.S. policy and management." *Ecological Applications* 16: 2035–2054.

Mack, R. N. 2003. "Plant naturalizations and invasions in the Eastern United States: 1634–1860." *Annals of the Missouri Botanical Garden* 90: 77–90.

Maniez, S. 2020. "Autumn olive guide + recipes." *Life's Little Sweets*, 5 September 2020. https://www.lifeslittlesweets.com/autumn-olive-guide-recipes/

Manuel, J. & J. Waller. 2021. "Coastwide nutria control program, 2020–2021." Coastal and Nongame Resources, Louisiana Department of Wildlife and Fisheries. https://nutria.com/wp-content/uploads/2020-21-CNCP-annual-report.pdf

Massachusetts Prohibited Plant List. 2022. Massachusetts Department of Agriculture and Resources. https://www.mass.gov/service-details/massachusetts-prohibited-plant-list

Masters, B. 2017. "Wild horses, wilder controversy." *National Geographic*, 6 February 2017. https://www.nationalgeographic.com/adventure/article/wild-horses-part-one

McConnaughey, J. 2017. "Bad beasts, good treats: Feral hog slaughterhouse takes off." *Associated Press News*, 26 April 2017. https://apnews.com/article/lifestyle-oddities-business-food-and-drink-7633a9184b0849b6a754e403417c94f2

McGraw, J. B. 2001. "Evidence for decline in stature of American ginseng plants from herbarium specimens." *Biological Conservation* 98: 25–32.

McLain, R. J., P. T. Hurley, M. R. Emery & M. R. Poe. 2013. "Gathering 'wild' food in the city: Rethinking the role of foraging in urban ecosystem planning and management." *Local Environment* 19: 220–240.

Mellish, J. M., A. Sumrall, T. A. Campbell, B. A. Collier, W. H. Neill, B. Higginbotham & R. R. Lopez. 2014. "Simulating potential population growth of wild pig, *Sus scrofa*, in Texas." *Southeastern Naturalist* 13: 367–376.

Meredith, L. 2020. "Early spring foraging: Garlic mustard." *Mother Earth News*, 19 June 2020. https://www.motherearthnews.com/real-food/early-spring-foraging-garlic-mustard-zbcz1403/

Milanović, M, S. Knapp, P. Pyšek & I. Kühn. 2020. "Linking traits of invasive plants with ecosystem services and disservices." *Ecosystem Services* 42: 101072.

Mishan, L. 2020. "When invasive species become the meal." *New York Times Style Magazine*, 2 October 2020, https://www.nytimes.com/2020/10/02/t-magazine/eating-invasive-species.html

Moorhouse, T. P. & D. W. Macdonald. 2011. "Immigration rates of signal crayfish (*Pacifastacus leniusculus*) in response to manual control measures." *Freshwater Biology* 56: 993–1001.

Morgan, M. 2015. "From readers: Eat Mo Carp tries to keep Asian carp from rivers and onto more menus." *Columbia Missourian*, 1 December 2015. https://www.columbiamissourian.com/from_readers/from-readers-eat-mo-carp-tries-to-keep-asian-carp/article_a3ceacd0-7798-11e5-b381-d7ebf7310461.html

Morris, Jr., J. A., K. W. Shertzer & J. A. Rice. 2011. "A stage-based matrix population model of invasive lionfish with implications for control." *Biological Invasions* 13: 7–12.

Morrison, S. A., A. J. DeNicola, K. Walker, D. Dewey, L. Laughrin, R. Wolstenholme & N. Macdonald. 2016. "An irruption interrupted: Eradication of wild turkeys *Meleagris gallopavo* from Santa Cruz Island, California." *Oryx* 50: 121–127.

Mosby, H. S. 1949. "The present status and the future outlook of the Eastern and Florida Wild Turkeys." Transactions of the Fourteenth North American Wildlife Conference. Wildlife Management Institute, Washington D.C.

Mosby, H. S. 1975. "The status of wild turkey in 1974." Proceedings of the National Wild Turkey Symposium 3: 22–26.

Murawski, S. A. 2010. "Rebuilding depleted fish stocks: The good, the bad, and, mostly, the ugly." *ICES Journal of Marine Science* 67: 1830–1840.

Naylor, R. L., S. L. Williams & D. R. Strong. 2001. "Aquaculture—A gateway for exotic species." *Science* 294: 1655–1656.

Norris, K. A. 2018. "A review of contemporary U.S. wild horse and burro management policies relative to desired management outcomes." *Human-Wildlife Interactions* 12: 18–30.

Noyes, L. 2020. "Foraging for autumn olive berries & 11 recipes to make." *Rural Sprout*, 24 September 2020. https://www.ruralsprout.com/autumn-olive-berries/

Nuñez, M. A. & D. Simberloff. 2005. "Invasive species and the cultural keystone species concept." *Ecology and Society* 10(1): r4. https://ecologyandsociety.org/vol10/iss1/resp4/

Nuñez, M. A., S. Kuebbing, R. D. Dimarco & D. Simberloff. 2012. "Invasive species: To eat or not to eat, that is the question." *Conservation Letters* 5: 334–341.

Nuñez, M. A., R. D. Dimarco & D. Simberloff. 2018. "Why some exotic species are deeply integrated into local cultures while others are reviled." Pp. 21–48 *in* R. Rozzi, R. H. May Jr., F. S. Chapin III, F. Massardo, M. C. Gavin, I. J. Klaver, A. Pauchard, M. A. Nuñez & D. Simberloff (editors), *Biocultural Homogenization to Biocultural Conservation*. Springer Nature, Cham.

Olsson, K., W. Granéli, J. Ripa & P. Nyström. 2010. "Fluctuations in harvest of native and introduced crayfish are driven by temperature and population density in previous years." *Canadian Journal of Fishery Aquatic Science* 67: 157–164.

Pardini, E. A., J. M. Drake, J. M. Chase & T. M. Knight. 2009. "Complex population dynamics and control of the invasive biennial *Alliaria petiolata* (garlic mustard)." *Ecological Applications* 19: 387–397.

Parola, P. 2022. *Can't Beat 'Em, Eat 'Em* (blog). http://www.cantbeatemeatem.us

Pasko, S. & J. Goldberg. 2014. "Review of harvest incentives to control invasive species." *Management of Biological Invasions* 5: 21–25.

Pearson, D. E., Y. K. Ortega, J. B. Runyon & J. L. Butler. 2016. "Secondary invasion: The bane of weed management." *Biological Conservation* 197: 8–17.

Preston, M. 2016. "Meet the massively destructive garden weed that 'Tastes like rain.'" *Bon Appétit*, 31 May 2016. https://www.bonappetit.com/test-kitchen/ingredients/article/japanese-knotweed-recipes

Pugliese, G. 2010. "Asian carp gets a new name, 'Kentucky Tuna.'" *Organic Authority*, 8 July 2010. https://www.organicauthority.com/buzz-news/asian-carp-gets-a-new-name-kentucky-tuna

Ricciardi, A., T. M. Blackburn, J. T. Carlton, J. T. A. Dick, P. E. Hulme, J. C. Iacarella, J. M. Jeschke, et al. 2017. "Invasion science: A horizon scan of emerging challenges and opportunities." *Trends in Ecology & Evolution* 32: 464–474.

Robbins, T. 1980. *Still Life With Woodpecker*. Bantam Books, New York.

Roman, J. 2004. "Eat the invaders!" *Audubon Magazine*, October 2004.

Rose, L. M. 2016. "Eat the weeds: Strawberry-knotweed pie." *Burdock & Rose*, 11 May 2016. https://burdockandrose.com/2016/05/11/japaneseknotweed-pie/

Rose, S. 2020. "Foraging for garlic mustard and how to use it." *Vintage Kitchen Vixen*, 5 May 2020. https://vintagekitchenvixen.com/foraging-for-garlic-mustard-and-how-to-use-it/

Sasser, C. E., G. O. Holm, E. Evers-Hebert & G. P. Shaffer. 2018. "The nutria in Louisiana: A current and historical perspective." Pp. 39–60 *in* J. W. Day & J. A. Erdman (editors), *Mississippi Delta Restoration*. Springer Nature, Cham.

Scasta, J. D., J. D. Hennig & J. L. Beck. 2018. "Framing contemporary U.S. wild horse and burro management processes in a dynamic ecological, sociological, and political environment." *Human-Wildlife Interactions* 12: 31–45.

Schröder, A., A. van Leeuwen & T. C. Cameron. 2014. "When less is more: Positive population-level effects of mortality." *Trends in Ecology & Evolution* 29: 614–624.

Seaman, A. N., A. Franzidis, H. Samuelson & S. Ivy. 2021. "Eating invasives: Chefs as an avenue to control through consumption." *Food, Culture & Society* 25: 108–125.

Seastedt, T. R. 2015. "Biological control of invasive plant species: A reassessment for the Anthropocene." *New Phytologist* 205: 490–502.

Simberloff, D. 2012. "Risks of biological control for conservation purposes." *BioControl* 57: 263–276.

Smith, N. S., S. J. Green, J. L. Akins, S. Miller & I. M. Côté. 2017. "Density-dependent colonization and natural disturbance limit the effectiveness of invasive lionfish culling efforts." *Biological Invasions* 19: 2385–2399.

Solomon, R. 2021. "Kudzu is so much more than the 'Vine that ate the South.'" *Slate Magazine*, 28 August 2021. https://slate.com/news-and-politics/2021/08/kudzu-south-japan-metaphor.html

Stauffer, T. 2017. "Autumn olives—The invasive super berry." *Northeast Superfooods*, 4 October 2017. https://www.northeastsuperfoods.com/blog/2017/10/4/autumn-olives-the-invasive-superberry

Storer, T. I. 1922. "The eastern bullfrog in California." *California Fish and Game* 8: 219–224.

Tabak, M. A., A. J. Piaggio, R. S. Miller, A. Sweitzer & H. B. Ernest. 2017. "Anthropogenic factors predict movement of an invasive species." *Ecosphere* 8: e01844.

Taylor, M. S. 2007. "Buffalo hunt: International trade and the virtual extinction of the North American bison." National Bureau of Economic Research (NBER) Working Paper No. 12969. https://www.nber.org/papers/w12969

Tsehaye, I., M. Catalano, G. Sass, D. Glover & B. Roth. 2013. "Prospects for fishery-induced collapse of invasive Asian carp in the Illinois River." *Fisheries* 38: 445–454.

Turner, B. C., C. E. de Rivera, E. D. Grosholz & G. M. Ruiz. 2016. "Assessing population increase as a possible outcome to management of invasive species." *Biological Invasions* 18: 533–548.

USDA APHIS. 2020. "Feral swine damage." https://www.aphis.usda.gov/aphis/ourfocus/wildlifedamage/operational-activities/feral-swine/feral-swine-damage

USDA APHIS. 2022. "History of feral swine in the Americas." https://www.aphis.usda.gov/aphis/ourfocus/wildlifedamage/operational-activities/feral-swine/sa-fs-history

van Kleunen, M., E. Weber & M. Fischer. 2010. "A meta-analysis of trait differences between invasive and non-invasive plant species." *Ecology Letters* 13: 235–245.

Vimercati, G., A. F. Probert, L. Volery, R. Bernardo-Madrid, S. Bertolino, V. Céspedes, F. Essl, et al. 2022. "The EICAT+ framework enables classifications of positive impacts of alien taxa on native biodiversity." *PLOS Biology* 20: e3001729.

Weber, M. J., M. J. Hennen & M. L. Brown. 2011. "Simulated population responses of common carp to commercial exploitation." *North American Journal of Fisheries Management* 31: 269–279.

Weber, M. J., M. J. Hennen, M. L. Brown, D. O. Lucchesi & T. R. St. Sauver. 2016. "Compensatory response of invasive common carp *Cyprinus carpio* to harvest." *Fisheries Research* 179: 168–178.

Wesserle, M. 2017. "Garlic mustard: A delicious invasive." *Four Season Foraging*, 28 May 2017. https://www.fourseasonforaging.com/blog/2017/5/28/garlic-mustard-a-delicious-invasive

White House. 2016. *Champions of Change*. https://obamawhitehouse.archives.gov/champions

Wilcove, D. S., D. Rothstein, J. Dubow, A. Phillips & E. Losos. 1998. "Quantifying threats to imperiled species in the United States." *Bioscience* 48: 607–615.

Zavaleta, E. S., R. J. Hobbs & H. A. Mooney. 2001. "Viewing invasive species removal in a whole-ecosystem context." *Trends in Ecology & Evolution* 16: 454–459.

CHAPTER 6

Turn Your Livestock into Weed Managers

Kathy Voth

In 2004, I developed a method to teach cows to eat weeds. In a first-of-its-kind pilot project, I taught a group of heifers to eat Canada thistle, leafy spurge, and spotted knapweed. Over the next couple of years, I modified the training process based on what I learned from my cow trainees as I worked with ranchers across the United States and Canada. I trained over 1,000 cows, flocks of sheep, goats, and a herd of bison on one of Ted Turner's ranches to eat a wide variety of weeds.

As it turns out, teaching a cow (or other livestock) to eat a weed is very easy. It takes just eight hours spread over seven days. Educated animals add other pasture weeds to their diets on their own and they teach their offspring and herd mates. They don't forget their new weeds over the winter, or ever, so you only need to train one group of animals once, and you're done. Weeds are also very nutritious so animals gain weight at the same rates or better than their grass-grazing counterparts. Best of all, graziers with an educated, weed-eating herd find they have about 43% more forage in their pastures, and they no longer have to buy or apply expensive herbicides.

With all that going for it, I thought that farmers and ranchers would be happy to adopt this practice. But changing minds has proven difficult because we've thought of weeds as our enemies for so long. So, in this chapter, I'll explain the science behind the training process, including what we know about how animals choose what to eat, lay out the training process and some cautions, and answer frequently asked questions. I've also included a short list of edible and inedible plants, some videos demonstrating some of the science, and additional plant resources. I hope it helps you, or graziers you know, to make life easier and more profitable.

It Started with a Goat

In 1996, I was working for the Bureau of Land Management. As part of my duties, I was acquainting our local community with the resources we managed by taking folks on weekend hikes. Participants always showed up unprepared—not enough water, no sunscreen, no snacks—so my pack was always heavy with extra supplies. To make my life easier and more interesting I bought a pack animal: a goat named Cisco.

From one goat, I went to two, and then 15, and then 50 and then 110—a grazing workforce as part of a seven-year research project on building firebreaks to slow and stop fires. I went from looking at grass as forage to seeing everything as potential food. And, like many crazy goat people, I began extolling goat virtues to the cattle ranching community. Why, there was even research published by Jean-Marie Luginbuhl and colleagues showing that adding two goats for every cow in a pasture would reduce weeds and improve the quantity and quality of forage for both goats and cows (Luginbuhl et al., 1999).[1]

The ranchers were not persuaded. They countered that goats are more susceptible to predators and they require different fencing and marketing so adding them to a ranching landscape just didn't make sense.

But graziers still had a weed problem, so I decided to modify my approach. Colleagues at Utah State University studying how animals choose what to eat found that they learn what to eat and choose diets based on nutrients and toxins in the foods. "Well," I thought. "If all that's true, then I should be able to teach a cow to eat a weed, and that would be a lot easier than getting ranchers to buy goats."

That was the inspiration for my first pilot project. I based my training process on research by Dr. Fred Provenza and his colleagues (https://extension.usu.edu/behave/), and on decades-old science describing how animals learn. Here's what I considered.

---

[1]. You can read more about Luginbuhl's recommendations for grazing with cattle and goats together here: https://content.ces.ncsu.edu/co-grazing-meat-goats-and-beef-cattle-has-many-advantages

Animals Learn What to Eat

One of the most important discoveries made by Provenza and his colleagues is that animals learn what to eat.

They learn first from their mothers. To illustrate this, I usually show a video of two different groups of ewes and their young. The first group of ewes only eat a plant called caragana, and when their lambs are put into the pen and given a choice between caragana and Russian olive, the lambs choose caragana. The second group of ewes only eats Russian olive, and, as the video shows, their lambs choose Russian olive over caragana when they enter the pen by themselves. Finally, when the two groups of lambs are put in the pen together with the two plants, they separate into their groups according to what their moms ate. (The link to the video is in the "Videos" section of this chapter.)

Researchers have found that these kinds of preferences passed from mom to offspring are very long-lasting, and even when a young animal learns to eat new things, it will always show a preference for the foods it learned to eat from its mom.

This is also why animals don't eat weeds on their own and we need to teach them. As long as they have enough familiar food to eat, they just won't try anything that hasn't been eaten by someone they know. In addition, since we have always considered grass to be good and weeds to be bad, we've often managed our pastures and livestock in ways that allow them to eat only the things they're most familiar with.

Palatability

The reason I had hope for changing a cow's mind about weeds was that animals also learn from experience. That's where palatability comes into play.

Most of us grew up believing that palatability is linked to the taste of a food. But Dr. Provenza's research shows that taste is simply a reflection of the body's experience of a food. Here's how it works: *all plants contain both nutrients and toxins.*

Nutrients generally send good feedback to the brain, and the brain links that positive experience to the flavor of the food. Nutritious foods taste "good" and the animal will tend to eat more of that food (Fig. 6-1). Foods low in nutrients provide less positive feedback, so they don't taste as "good." That's why high-nutrient, high-calorie ice cream tastes so much better than low-nutrient, low-calorie celery. (I've included a link to "Sheep who love straw," a video showing how animals respond to a nutrient-rich food vs. a poor-quality food.)

TURN YOUR LIVESTOCK INTO WEED MANAGERS 115

Meanwhile, toxins provide negative feedback, normally experienced as nausea. The higher the dose of a toxin, the more negative the feedback and the less the animal eats. The brain links the taste of the food to the negative feedback and labels that flavor "bad." (Check "Videos" for a video showing how animals respond to toxins in forages.)

That meant my weeds had to be nutritious to be delicious. As I found out, they are. *Most are equal to or better than grass in quality and also low in toxins.*

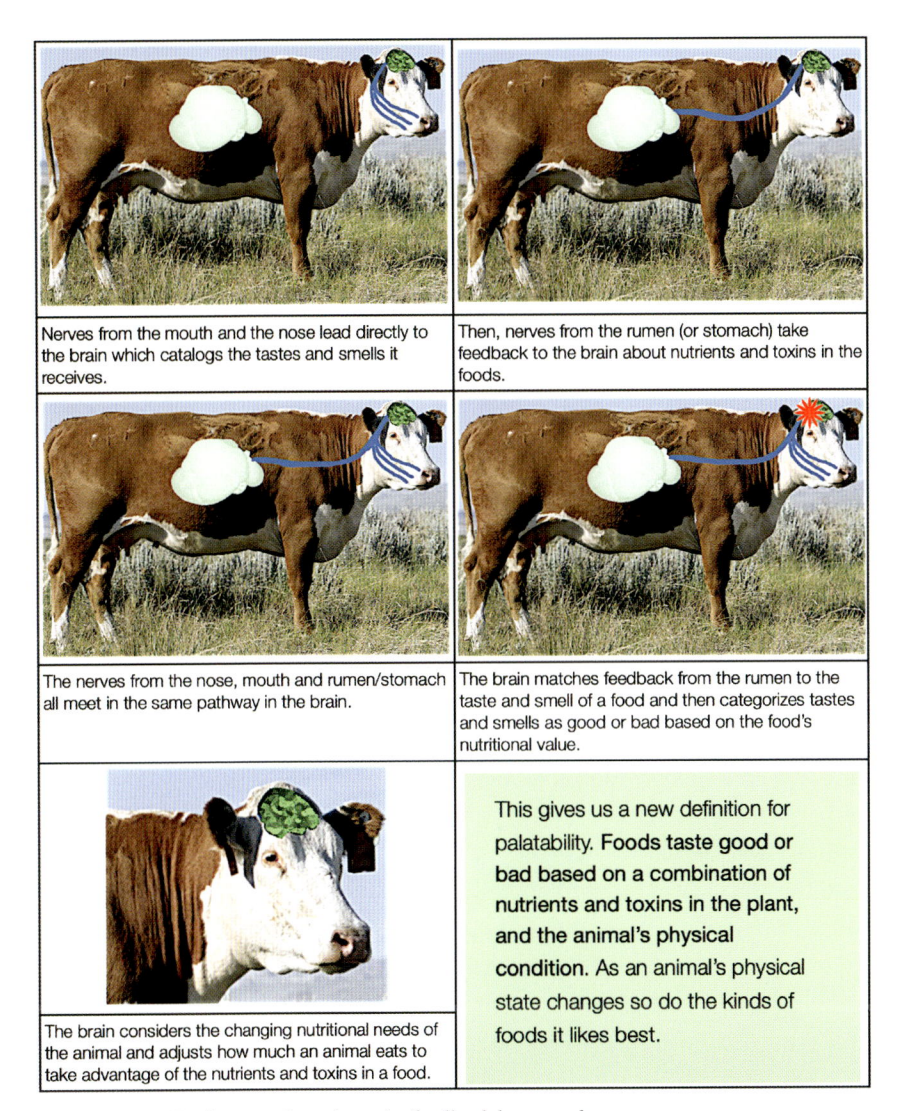

Nerves from the mouth and the nose lead directly to the brain which catalogs the tastes and smells it receives.	Then, nerves from the rumen (or stomach) take feedback to the brain about nutrients and toxins in the foods.
The nerves from the nose, mouth and rumen/stomach all meet in the same pathway in the brain.	The brain matches feedback from the rumen to the taste and smell of a food and then categorizes tastes and smells as good or bad based on the food's nutritional value.
The brain considers the changing nutritional needs of the animal and adjusts how much an animal eats to take advantage of the nutrients and toxins in a food.	This gives us a new definition for palatability. **Foods taste good or bad based on a combination of nutrients and toxins in the plant, and the animal's physical condition. As an animal's physical state changes so do the kinds of foods it likes best.**

FIGURE 6-1. Here's more about how the feedback loop works.

The Fear of New Foods

Now, here's the real problem: how to get a cow to take a bite of something she's never tried before and has never seen anyone eat.

To different degrees, we're all afraid of trying new things. But when I looked at animal behavior research, I learned that the more positive experiences animals have with new things, the more likely they are to try other new things. Along with that, I thought about Pavlov's dog[2] who learned from routine that every time the bell rang, he was going to get fed. I combined these two ideas to create a routine where animals tried so many new things that new became normal and not scary.

Here's what the routine looks like: I drive up in the same truck, at the same time twice a day, and put an unfamiliar but nutritious snack in the same tubs every time. I do this for four days, giving the trainees eight different foods to try. Why eight? After a close look at data from Provenza's experiments, I could see that seven to eight experiences were what it took for animals to be comfortable with something new.

Trainees are quick to learn the routine, and they figure out that every time I show up, I'm going to give them something tasty. As I watched how they behaved, I also learned that peer pressure could help me get where I was going. By using one tub for every three animals, they were quicker to grab what they could get before the next cow could eat it all.

On day five, I can introduce the weed—just another new food in a series of new things (Fig. 6-2). They try it like they've tried everything else, get the good feedback from the nutrients, and they're ready for the next step—eating it in pasture on their own.

The Training Process

Now let's take a closer look at all the steps in the training process.

1. Know Your Weed

With this step, we're focused on making sure our weed is reasonably high in nutrients and that whatever toxins it contains aren't harmful.

---

[2]. Ivan Pavlov was a Russian psychologist. His discovery of classical conditioning (another term for routine) began accidentally. He was studying dogs' salivation response to being fed and noticed that they began salivating when they heard the footsteps of the person coming to feed them. The resulting experiment involved ringing a bell before the dog was fed, teaching the dog to associate the bell with upcoming food. In the end, he could ring the bell and the dog would drool.

FIGURE 6-2. A Hereford heifer learns to eat leafy spurge. (Photo by author.)

Palatability/Nutrition

Finding the nutritional value of weeds is not easy. Resources available on the internet typically conclude that weeds are not tasty because animals don't eat them. But, as we've already learned, that's not a good measure of whether or not a food is good. The animal might be avoiding it simply because it's unfamiliar. You can send it in for testing, as I did with many weeds when I first began training livestock. But that can be time-consuming, and over time the expense adds up.

Fortunately, the work I did testing weeds led to a good rule of thumb: if something is green and growing, it's generally nutritious, and the higher the leaf-to-stem ratio it has the more nutritious it will be. As plants mature or dry, their nutritional value drops, making them less palatable.

Toxin Content Determines What You Can Train Your Livestock to Eat

Just like nutritional value, it can be hard to find information on what weeds are not safe for your livestock to eat. Though most toxins only reduce the amount an animal will eat, some toxins cause death, abortion, birth defects, or other serious health problems. *Don't skip this step and NEVER try to train an animal to eat a weed that could cause harm.* At a minimum, you'll put your animals off trying things you bring them, but you also risk their health.

It's the Dose That Matters

Just because a plant contains a toxin does not mean it is toxic. In addition, research shows that animals can safely mix their own diet based on the feedback from the forages they eat. As long as they have plenty of variety to choose from, they are unlikely to overeat one food to a deadly dose. But pay attention to the "Cautions" section. Remember that most poisonings are a result of human error. Putting animals in a pasture with little variety or putting very hungry animals in a new pasture can result in them overeating a dangerous food. We see this most often in cases of bloat in alfalfa pastures.

Check the "Edible/Nonedible Weeds" section for information on species of concern and "Weed Information Sources" for a link to a booklet I put together with information on all the plants I've trained animals to eat or looked up for other people. I've also included information on the books I use to find information about toxins in weeds.

2. Choose Your Trainees

What I've learned from over a decade of training is that any animal can learn to eat a new food. I started with yearling heifers because research and experience show us that young animals are more likely to try new things. But when a rancher didn't have heifers, I learned that cow-calf pairs worked just as well. Old cows were no slower, bulls learned quickly, and all the sheep, bison, and goats learned to eat new weeds.

My whole goal when developing the training method was to make it easy for the people involved. So, when you're choosing your trainees, choose what's easiest for you. Just be sure that you're working with healthy animals that will be on your place for a while, and that they are in a convenient pasture. I also like to choose a pasture that has my target weed in it so that once the trainees have tried the weed in the tubs, they can immediately try it in pasture.

How many animals you'll train depends on your herd size and circumstances. But here are some things to remember:

- We'll be using our animals' competitive nature as part of the training, so it's helpful to work with a group of at least 10 or more. The reason for this is that I've learned that when we work with fewer than a dozen animals, we are more likely to see them as individuals and then we treat them differently. That tends to reduce competition among animals and makes training more difficult.

- I generally train groups of 25 to 50 at a time. It's easier than training a whole herd and achieves the same impact. Based on a decade of experience I know that my trainees will teach their offspring and herdmates and that it doesn't take very many weed-eating animals to turn an entire herd into weed-eaters. I've seen 12 teach 110, steers learn from trained heifers in the pasture next to them, and 50 cows learn from a herd of 38 weed-eaters in just one day.

3. Make the Unfamiliar Seem Familiar

Routine makes everything seem normal. Your job is to show up at a time that's convenient for you in the morning and afternoon, making the same sounds, driving the same vehicle, with the same feed tubs/troughs, and something tasty for your trainees to try. (If twice a day doesn't work for you, adjust the training to suit you. Just be sure that you maintain a routine.)

Tasty Snacks

You'll feed an unfamiliar nutritious food every morning and afternoon for four days. Just go to the feed store and pick eight different things, one 50-pound bag per 25 cattle, plus one extra bag that you'll use to mix with the weeds later. Remember, you're going to show them that food comes in a variety of shapes and sizes, so when you're choosing your snacks, pick a variety of textures, flavors, shapes, and smells (soy flake, wheat bran, rolled oats, alfalfa pellets, range cubes, and COB, for example).

The routine teaches trainees that every time you show up, they may not recognize what you're giving them, but it will surely be something good to eat. When you serve up weeds, they will be just one more new thing in a series of strange things.

Teaching Tools

I add *sight and sound cues* to the training as well. The sight and sound of my truck driving to the feeding site and a honking horn are all good ways to announce you're coming, call the animals in, and create excitement before you arrive.

Feed tubs are also helpful. In the West, graziers often feed supplement that comes in large tubs. I like them because their shape and size encourage competition and let me use the cow's natural behavior to help them learn what to eat. The tubs are big enough that three heifers or two cows can stick their heads in at the same time. The sides are tall enough that a trainee can't see what her herd mates are eating. She just seems to assume that since everyone else is eating, there must be something good, so she shoves her head in and grabs what she can get. This level of competition helps us because animals put things in their mouths without thinking so much. Since we're giving them nutritious foods, we know that we've won the battle once a food makes it into the gut.

If you don't have your own supplement tubs, check with neighbors or your feed store. I've found them waiting in recycling and I can use them and return them. You can also use other kinds of feeders, just choose something that has the tub attributes for competition.

If you watch the video of the training process in action, you'll notice that I drizzled a bit of molasses onto the weeds. I initially used molasses in training for two reasons. First, research showed that a familiar flavor carried from one food to the next helped animals try new things. It's the same as me offering you frog legs and getting you to try them by telling you "they taste like chicken." Since western ranchers often feed a molasses-based supplement to cattle, it's a familiar flavor. Molasses also acted as a carbohydrate to help animals deal with nitrates in some weeds. You'll find examples of nitrate-containing plants in the "Edible/Nonedible Weeds" list.

People mistakenly assumed that molasses was the magic behind the success of the project and thought that they could shortcut the training by simply spraying molasses on their pastures. This has limited success in most cases. To explain why, I ran a simple trial.

I took two groups of animals. One had been through the training up to the point where they were going to try weeds for the first time. The other was untrained. I offered each group a mixture of molasses, grain, and weeds. The untrained animals ate 3 ounces, the trained animals ate 81 ounces.

This demonstrated that it is not the molasses that makes the difference, and in fact, the only time I add molasses is when someone I'm working with is feeling nervous about feeding weeds. These nerves are picked up by the trainees who become reluctant to try the weeds because the person feeding them seems unsure. Drizzling a little molasses on the weeds calms humans and the trainees are happy to eat.

Breaking the Routine

On days five through seven, we break the routine and skip the morning feeding. This does two things. First, it makes trainees a little anxious so they are more excited to try whatever is in the tubs that afternoon. It's also a good behavioral technique for ensuring that animals show up whenever you arrive because they're hoping you'll be bringing a snack.

Then at the afternoon feeding, clip weeds, mix them with a bit of feed, and serve them up. How many weeds do you need? Loosely fill two 50-pound grain sacks or one 250-pound supplement tub per 25 cattle trainees. I repeat this on day six, and on day seven I feed the weeds by themselves.

If you'd like to see the process in action, please check "Videos" where you'll find a link to my YouTube channel and videos you can watch.

4. Observe Your Trainees in Pasture

By the second or third day of weed feeding, I find that the trainees have discovered the weeds in their pastures and are trying to eat them. I stop as soon as I see this, and I never feed weeds for more than three days. If you're worried that they haven't seen the weeds or want to make sure that you can see them trying them, here are some things you can do:

- Place a feed tub near a patch of weeds. This will attract your animals to the patch of weeds. A salt block or supplement block works too. While they're waiting for their turn, trainees will graze whatever is in the vicinity, including the weeds.

- Sit near a patch of weeds in the evening when your trainees are grazing. Since you've been bringing them treats, they'll come to check you out. If you just sit quietly, they'll get bored, begin grazing, and will probably choose some of the weeds.

A weed's growth form differs from the grasses your trainees are accustomed to grazing. That means they need to learn new techniques. Watch for changes to the target weeds that show they are trying. Bent stems or stems that are bitten off are all indications that they are trying.

Don't expect them to clear the pasture of weeds or to eat only weeds. They will continue to mix forages for a healthy diet that meets their individual needs.

Cautions

You are responsible for the health and safety of your animals. Here are rules to follow to keep them that way.

Never Put Animals in a Solid Stand of Any Plant

All plants contain toxins. It is the dose that can cause illness or death. A variety of forages in pasture can prevent overdosing on any one thing.

Let Animals Start Slowly to Adjust to New Foods

It takes time for rumen microbes to adjust to new plants. Ruminant nutritionists typically recommend 10 days to two weeks for microbial populations that degrade secondary compounds to increase. The gradual increase in intake over seven to 10 days by animals naïve to a particular plant enables microbial populations that degrade compounds in the plant to gradually increase in abundance.

Don't Put Hungry Animals into a New Pasture

Think of what happens when you're hungry and grocery shopping. You tend to buy things you shouldn't. Hungry, thirsty animals are just like that. They don't make good choices.

This rule is especially true for pastures largely made up of nitrate-accumulating plants. Make sure that they have a full stomach when moving to a new pasture with large stands of things like Canada thistle or pigweed. This will ensure that they have the necessary carbohydrates in their rumen to aid the rumen microbes in breaking down nitrate.

The problem with nitrate accumulators is not the nitrate itself, but the chemical process that takes place in the rumen. Nitrate is reduced in the rumen in a series of steps from nitrite to nitrate to ammonia and finally to microbial proteins. It is the nitrite in this step that causes poisoning if it accumulates in large amounts. When the animal has carbohydrates in its rumen, rumen microbes are better able to convert nitrates to ammonia and proteins, preventing the potential for nitrite accumulation. Researchers have also found that rumen microbes can adapt to nitrate processing over a period of five to seven days of slow introduction to nitrate-accumulating plants (Knight & Walter, 2001). So the best thing to do, when working with ruminants on nitrate-accumulating plants like goldenrods and Canada thistle, is to allow the rumen microbes time to adjust to nitrates in plants by feeding just a little at a time and providing plenty of variety in pasture.

Canada thistle is the reason my original training process included molasses. According to Knight and Walter, "when carbohydrates such as corn and molasses are present in the rumen, nitrates are more rapidly converted to ammonia and microbial proteins without the accumulation of nitrite" (Knight & Walter, 2001). When I first began feeding Canada thistle to my trainees, I mixed it with corn and a molasses-based vitamin product. Since they were familiar with corn and molasses, I knew it would encourage them to try the weed, while at the same time providing them with important protection. I later found that I was feeding such small quantities that neither the molasses nor the corn were necessary for the rumen, though the molasses was helpful when it came to calming the nerves of ranchers concerned that their trainees couldn't learn to eat weeds.

Answers to Frequently Asked Questions

Do I have to train livestock to eat every weed in pasture?
No. What I've learned is that animals become more willing to try things in pasture after going through the training process, and learning to eat a new plant, they will try other things they find in pasture. If they are not trying something I know is safe for them to eat, I bring back a training tub, cut some of the weed, and put it in the tub. Because trainees learned that everything you put in the tub is good food, they will try it, and then begin to eat it in pasture. This is something the cows taught me and I've done it lots of times.

Will weed-eating livestock eliminate weeds from pasture?
Probably not. I do know of one instance where cows eliminated Italian thistle from their pastures, but it's more normal that they will simply manage the weeds so they are not a problem. Just remember, these are no longer weeds, they are forage. My review of literature has led me to estimate that for an average pasture, over 40% more forage will be available if animals graze weeds rather than avoiding them, though this number varies depending upon what plants are present.

Can weed seeds be spread by manure?
Yes, and they can also be spread when the seeds stick to their bodies. Weed seed spread is something that is constantly happening via animals, people, equipment, wind, and rain. This is less of a concern when your livestock eat the plants that grow from the weeds.

Declare Peace in the War on Pasture Weeds

While this solution doesn't work in all cropping systems, in pasture, it makes good sense to turn weeds into the nutritious, resilient forage they can be. Use this simple process to turn your livestock into weed eaters, and make your grazing life easy and more profitable.

Edible/Nonedible Weeds

Here is a short list of the weeds in the eastern United States that I am most often asked about. I provide a complete list in the ebook *Edible Weeds & Training Recipe*. Information for that book is available below.

Autumn Olive (*Elaeagnus umbellata*)

Missouri rancher Greg Judy loves this tree as a forage for his cattle. The berries are also very tasty and high in antioxidants. By cutting the trees off at the stump, the tree resprouts with lots of tender shoots that are great for grazing. The tree is also a nitrogen fixer, making it beneficial to other forages in your pastures. You can read more here: https://onpasture.com/?s=Autumn+olive&post_type=post.

Canada Thistle (*Cirsium arvense*)

This is one of the easiest weeds to teach livestock to eat because it compares to alfalfa in nutritional value (21% protein in spring, 13% in summer, and 12% in fall). If you have this in your pasture, I'd highly recommend starting here. Once animals are eating this thistle, they quickly add other thistles to their diet and then begin looking around to see what other kinds of plants in the pasture could make good forage (Fig. 6-3).

FIGURE 6-3. A group of trained heifers eating Canada thistle in Lewiston, Montana. (Photo by author.)

Spines are of no concern at all to grazers. It's actually the nitrates in this plant that cause a little bit of concern. To keep animals safe:

- Give rumen microbes time to adjust, introducing the food in small amounts over five to seven days.

- Never put animals in a solid stand of this plant.

- Don't put hungry animals into a field that is primarily Canada thistle. Full rumens prevent nitrate poisoning.

Other Thistles: Musk or Nodding (*Carduus nutans*), Bull (*Cirsium vulgare*), Scotch (*Onopordum acanthium*)

All thistles are very edible including these. Train them to eat one kind of thistle, and your cattle will eat them all.

Common Buckthorn (*Rhamnus* spp.)

The seeds and, to a lesser extent, the leaves have laxative properties. But, because such a large amount has to be eaten, this plant is not a problem. I might teach cattle to eat something else first, and then let them add this to their diet later.

Crown Vetch (*Coronilla varia*)

Toxic Plants of North America describes this as "a palatable and useful forage of somewhat lesser value than alfalfa" adding that its commercial use was curtailed because of concerns about the presence of cardiotoxins and nitrotoxins, particularly for horses (Burrows & Tyrl, 2013). Some other vetches are more toxic, but in prolonged feeding trials of this species, researchers were unable to produce adverse effects in ruminants who eat the plant readily. This is because the primary toxin, 3-NPA, is detoxified in their rumens. If you are concerned, train them to eat other foods first and then let them choose this on their own.

Garlic Mustard (*Alliaria petiolata*)

I trained cows in California at Vandenberg Air Force Base to eat several types of wild mustard. They ate it happily (see "Videos" below). It is similar to hoary cress in nutritional value—30% protein when bolting to 8% in full seed. As a member of the *Brassica* family, it contains glucosinolates, which can cause enlarged thyroids. Fortunately, ruminants are not too prone to thyroid issues unless they eat very large quantities of *Brassica* plants.

Heavenly Bamboo (*Nandina domestica*)
Do not train your animals to eat this. Many but not all cultivars are strongly cyanogenic and are a hazard to animals that gain access to them.

Honeysuckle: Bush (*Lonicera tatarica*) and Japanese (*Lonicera japonica*)
Deer regularly eat this plant as do some livestock. It maintains year-round crude protein of between 12% and 16%, making it a very palatable choice.

Horse Nettle (*Lythrum salicaria*)
I worked with Don Ashford in Louisiana to teach his herd to eat this plant in pasture. They have been eating it for more than five years now with no harmful effects. You can read more here: https://onpasture.com/2020/04/20/yes-your-livestock-can-eat-horsenettle-and-some-other-southern-favorites-too/.

Japanese Knotweed (*Fallopia japonica*)
This is a good forage, and people can eat it too! Start ruminants slowly. Knotweeds can be nitrate accumulators, so rumen microbes need a week or more to adapt. Here's an article with a link to a five-minute podcast about eating it, with some recipes too: https://onpasture.com/2013/03/19/the-tart-taste-of-knotweed/.

Kudzu (*Pueraria montana* var. *lobata*)
This plant is comparable to alfalfa in nutritional value and should be grazed early and often to keep it under control. The flowers can be made into a jelly that tastes like grape jelly, and when bees use it they turn it into a runny red or purple honey that tastes like grape jelly or bubblegum.

Leafy Spurge (*Euphorbia esula*)
It is not true that the latex in this plant causes harm to cattle. This is one of the first plants I trained cattle to eat in Montana. I chose it because I knew of a herd in Nebraska that ate this plant, and thought if they could do it, I should be able to teach other cattle to eat it. Trainees grazed it in pasture, doing better when their pastures contained more variety. When they only had a little grass and lots of spurge, they didn't graze as much spurge.

Multiflora Rose (*Rosa multiflora*)
This is another great plant to graze with cattle. Don't worry about the thorns. Cattle don't seem to mind them on this plant, or any other. There are no toxins

of concern and it runs about 12%–15% protein depending on how much of the woody part the animal eats. When training animals to eat this plant, clip the softer, tender, leafy ends of branches. It might take a bit longer to harvest than other weeds, but the end result is worth it. A farmer I worked with in West Virginia is just tickled by how much multiflora rose his cattle eat.

Perilla Mint (*Perilla frutescens*)

This plant is toxic. Do not teach your animals to eat it.

Sericea Lespedeza (*Lespedeza cuneata*)

Society for Range Management articles from the '40s and '50s talk about animals eating this and one of our On Pasture authors grew it as hay in the '60s. I highly recommend teaching cattle to eat this plant. That said, there are still some mysteries to be solved about encouraging livestock to eat it thanks to the tannins it contains. Researchers at Utah State University found that polyethylene glycol can be used to increase intake as shown in this paper: https://repository.arizona.edu/handle/10150/639955. A reviewer of this chapter also is aware of a rancher who had very good results grazing a combination of sericea and endophyte-infected tall fescue, evidently because the tannins in sericea bind with the alkaloids in fescue, rendering both compounds less active.

Smartweed (*Polygonum* spp.)

Do not teach your animals to eat this. Although experiments on this species' toxicity have resulted in only suspicion, when problems do occur, animals have died. Chemicals in the plant may cause sensitivity to sunlight.

Spotted Knapweed (*Centaurea maculosa*)

My first training attempt in 2004 included Canada thistle, leafy spurge, and spotted knapweed. Since then, I've trained lots of cattle to eat spotted knapweed, and, next to Canada thistle, it is one of my favorite weeds to train cows to eat. It is equal to alfalfa in nutritional value, animals take to it quickly, and they eat a lot in pasture (Fig. 6-4). If you graze spotted knapweed before seed set, it will put out new flowers. But research has shown that the majority of the seeds are not viable. Here is a paper I wrote on managing grazing to reduce spotted knapweed populations: https://onpasture.com/2020/04/06/managing-livestock-to-meet-weed-management-goals/. And here's Jim Gerrish writing about how he did it with his own cattle: https://onpasture.com/2022/05/09/high-density-grazing-for-spotted-knapweed-suppression/.

FIGURE 6-4. Three trained cows eating diffuse knapweed in pasture in Boulder County, Colorado. (Photo by author.)

There are many varieties of knapweed. All are edible and respond similarly in pasture.

Videos

Animals Learn from Mother What to Eat: https://youtu.be/JybcA5fe_qs

When Good Lambs Go Bad (an explainer for why one lamb in the video behaves differently than expected): https://youtu.be/B6-6Tv4WmHc

Sheep Who Love Straw: https://youtu.be/3_07XrsoDCw

Animals Learn What to Eat from Experience (sheep respond to toxins in their food): https://youtu.be/ZUU9rlws5bs

The Training Process in Action—Cows Learning to Eat Weeds in Boulder County Colorado: https://youtu.be/wAUoNtOFbCE

Please note that the training site was located in a former coal mine area and was the only accessible location on a 500-acre pasture. While this part of the pasture is ugly, it had lots of great forage elsewhere.

Teaching Cows to Eat Black Mustard (cows at Vandenberg trained to eat wild mustard): https://www.youtube.com/watch?v=cwZUBQiuCCo

Additional training videos are available on my channel: https://www.youtube.com/user/KathyVoth.

More animal behavior information is available from Utah State University here: https://extension.usu.edu/behave/.

Weed Information Sources

Cows Eat Weeds/Edible Weeds & Training Recipe by Kathy Voth (ebook set)

This two-ebook set provides a more in-depth look at the science and a complete list of all the weeds I've either trained livestock to eat or have researched for others to train. You can find it here, along with links to other articles I've written on this topic: https://onpasture.com/2021/03/01/turn-your-livestock-into-weed-eaters/.

A Guide to Plant Poisoning of Animals in North America by Anthony P. Knight and Richard G. Walter, 2001, Teton NewMedia, Jackson, Wyoming (https://www.tetonnm.com).

The book includes pictures and information to help you identify plants, the principal toxins in them, clinical signs of poisoning, and treatment. Since the dose of a toxin is what causes poisoning, it also includes information on how much of a plant an animal has to eat to experience problems. You'll have to learn a little bit of new vocabulary to make use of everything in the book, but the effort is well worth it.

Toxic Plants of North America by George E. Burrows and Ronald J. Tyrl, 2001, Iowa State University Press, Ames.

This is a great reference book that covers all the toxins in all the plants in North America that we know of. It weighs in at about 8 pounds and is no longer available on Amazon, where I bought it for almost $200 in 2004. A second edition is now in print, published by John Wiley & Sons in 2013, which is even more expensive. It wasn't written for the layperson, but it is decipherable. One of the things I like about it is that it is a thorough review of a huge volume of research and information on toxins in plants.

Literature Cited

Burrows, G. E. & R. J. Tyrl. 2001. *Toxic Plants of North America*. Iowa State University Press, Ames.

Knight, A. P. & R. G. Walter. 2001. *A Guide to Plant Poisoning of Animals in North America*. Teton NewMedia, Jackson.

Luginbuhl, J.-M., T. E. Harvey, J. T. Green Jr., M. H. Poore & J. P. Mueller. 1999. "Use of goats as biological agents for the renovation of pastures in the Appalachian region of the United States." *Agroforestry Systems* 44: 241–252.

CHAPTER 7

Invasive Plants Used in Chinese Medicine

Thomas Avery Garran

THE PLANTS DESCRIBED in this chapter are all of Asian origin and are used in Chinese medicine. Chinese medicine is a complex medical paradigm with a living scholarly tradition dating back at least 2,000 years that has proven effective in many conditions via modern scientific methods; it is not discussed here. The plants that are available to us in North America can, perhaps should, be utilized both for self-care and as a source of economic benefit for those interested in commercial harvesting for sale to brokers or manufacturers.

This chapter is only meant to serve as an introduction to these plants. The information about their medicinal actions and traditional uses is based on the traditional Chinese literature and has been rendered in relatively simple language for the nonprofessional. Further, information on dosage and on other plants with which individual species are traditionally combined, which is essential to safe use of many plants, has not been included. Therefore, the information in this chapter is not meant to be a substitute for professional medical advice or services by licensed or otherwise experienced practitioners. Likewise, anyone intending to collect plants with which they are not already personally familiar is strongly advised to seek the assistance of a botanist, herbalist, or otherwise trained individual who can positively identify the plant of interest, prior to harvesting. While most of the plants in this chapter are relatively easy to identify, mistaken identity could lead to poisoning; *please be careful.*

Speaking of harvesting, while some plants have specific requirements, which are noted in the treatments of those plants, most harvesting of plant parts is relatively standard. The following is meant as an introduction, but proper training by an experienced wildcrafter is strongly recommended prior to gathering wild plants.

- Flowers: Flowers are gathered either during their bud stage or at their peak. Flowers should be harvested after any dew or rain has dried. Do not harvest wet flowers; they will rot easily and quickly.
- Leaves: Leaves are gathered during the day, after any dew has dried. If the leaves are aromatic, the best time to harvest them is around midday when they are the most fragrant.
- Roots and rhizomes: Roots and rhizomes are gathered in the autumn or spring. Most roots are best harvested from perennial plants in the autumn after the plant has died back for the winter.
- Bark: Bark from tree branches or roots is generally harvested in the spring around the time the tree's leaf-buds are swelling or just as they have started to leaf-out.
- Whole Plant: Sometimes the entire plant can be used. Generally, the more tender parts (i.e., newer growth) are considered the best quality, but in the case of annuals, the entire plant is usually harvested. Note that whole plant can sometimes include the root but not always. This will be noted for each plant specifically.

Each monograph provides readers with the most common name used, Latin or "scientific name," and the Chinese name using both the Romanized pinyin name and the traditional characters. Common names are important but they can be problematic because there are often many common names for a single species. The Latin or "scientific name" is the most specific and is only used for a single plant species. The Chinese names have been included for accuracy and completeness. Sources for descriptions, biogeographical notes, and uses include the *Flora of China* (Wu et al., 1994–2013), USDA Plants Database (USDA, NRCS, 2022), *Grand Dictionary of Chinese Medicinals* (Nanjing University of Chinese Medicine, 2009), and *Flora of Chinese Medicinal Plants* (Ye, 2017–2022).

Common Name: air potato

Latin Name: *Dioscorea bulbifera* L.
Chinese Name: huang yao zi 黃藥子

Part used: underground tuber

Description: The shiny stems sprout from the underground tuber. Twining to the left, they climb on just about anything. The smooth, heart-shaped leaves emerge alternately from the stem and are green to dark green with several longitudinal leaf veins. The plant has separate male and female flowers that look similar and produce light, papery, winged seeds. The plant also produces brown-purplish aerial tubers from the stem that can weigh up to 300 g.

Biogeographical notes: Primarily found in central and southern Florida but can be expected sporadically in other southern states. Usually found in disturbed areas, forest edges, and along streams and other bodies of water.

Common uses: This herb is used for scrofula, swollen and inflamed throat with inability to swallow, swollen and inflamed ulcers and sores, snake bite, tumors (has been used for some types of cancer), vomiting blood, nose bleed, coughing blood, whooping cough, and inflammatory lung ailments with coughing and wheezing.

Harvest time: Tubers are dug during the winter from plants that are at least two to three years old.

Drying and processing: The fibrous roots are cut off the tuber, then it is sliced in 1 cm thick slices and dried under forced air conditions.

Market potential: Low to medium. This is a relatively strong herb and only used occasionally in professional medicine in the United States.

Note: The tuber is said to have minor toxicity, although it is sometimes consumed. It must be cooked thoroughly before it can be consumed as a food.

Common Name: autumn olive

Latin Name: *Elaeagnus umbellata* Thunb.
Chinese Name: niu nai zi 牛奶子 (plant), niu nai zi ye 牛奶子葉 (leaf), niu nai zi gen 牛奶子根 (root)

Part used: ripe fruit, root, leaf

Description: This shrub has gray bark and, unlike its cousin *Elaeagnus pungens* Thunb. (thorny olive), is deciduous with only scattered spines. The gray-green leaves are 2–6 cm long and often have an acute (pointy) tip. The flowers are silvery-white and funnel-shaped, about 5–7 mm long. These give way to a multitude of small (6–9 mm long) red berries.

Biogeographical notes: This shrub is most common in disturbed areas, forest edges, and fields from southern New England, south to Virginia, and west to Indiana. It can be expected sporadically in other areas east of the Mississippi river as well as the Pacific Northwest.

Common uses: All the parts of the plant are used very similarly to treat lung inflammation with cough, inflammatory bowel conditions with diarrhea, urinary tract infections, vaginal discharge, breast infections, and excessive menstrual bleeding. The leaves are also used externally to treat sores and carbuncles.

Harvest time: See thorny olive (*Elaeagnus pungens*).

Harvest techniques: See thorny olive (*Elaeagnus pungens*).

Drying and processing: See thorny olive (*Elaeagnus pungens*).

Market potential: See thorny olive (*Elaeagnus pungens*).

Common Name: Bermuda grass
Latin Name: *Cynodon dactylon* (L.) Pers.
Chinese Name: gou ya gen 狗牙根

Part used: whole plant

Description: This grass creates thick mats of stoloniferous growth and reaches a height of approximately 40 cm at maturity. The green leaf blades grow from 1 to 5 cm long. It has yellow flowers that are borne from branches at the top.

Biogeographical notes: Bermuda grass is a well-known weed throughout much of the United States. It can be found on roadsides, in fields and meadows, and in cropland. It can live in most environments and is commonly found on disturbed ground.

Common uses: Bermuda grass is mainly used for stopping bleeding, including for coughing blood, nose bleed, and blood in the urine or stool due to physical exhaustion and chronic inflammation. It is also used for arthritic conditions with symptoms of swelling and stiffness of the joints. Externally, the fresh herb is pounded to a paste and applied to traumatic injuries with inflammation and red and inflamed sores and ulcers.

Harvest time: Best collected from July through September.

Harvest techniques: The entire grass with roots is harvested by simply pulling it from the ground.

Drying and processing: The herb is also made into a tincture and combined with other herbs to treat joint pain.

Market potential: Probably low but given how common this herb is, it could be useful for self-care.

Common Name: Chinese silvergrass
Latin Name: *Miscanthus sinensis* Andersson
Chinese Name: mang gen 芒根 (root), mang jing 芒茎 (rhizome)

Part used: root and rhizome

Description: Creating thick patches that can climb to 4 m tall, this ornamental grass has green leaves and a flowering head consistent with its name, silver. There are also several ornamental varieties. Aside from spreading via seed, it can also spread from rhizomes.

Biogeographical notes: Although present in many states east of the Mississippi river, the highest concentration of this invasive grass is in New Jersey, southern Virginia, and northern North Carolina. It is commonly found along the coast and prefers open slopes and disturbed places.

Common uses: Prepared in large doses as a decoction it is used to treat cough, scant and inhibited urination, thirst associated with fevers, and vaginal discharge. This herb is also used for absence of menstruation combined with emaciation, no desire to eat or drink, low-grade fever, and extremely dry skin.

Harvest time: Collected from August through November.

Harvest techniques: The grass is cut at the base with an appropriate tool.

Market potential: Low. Not commonly used but could be used for self-care and community health care.

Common Name: Chinese wisteria
Latin Name: *Wisteria sinensis* (Sims) DC.
Chinese Name: zi teng 紫藤

Part used: stem bark

Description: This well-known plant can grow up to 25 m long, twisting leftward around anything it can attach itself to. The young stems have abundant white hairs, giving them a near shiny appearance, eventually become woody, and can strangle other plants, including trees. The clusters of purple flowers are easy to recognize and very showy.

Biogeographical notes: The vining plant can be found from New England south and west to Florida and Texas. It is most commonly seen as an escaped weed in Louisiana and Alabama. It tends to be in areas disturbed by humans but can also be found at forest edges, hillsides, and roadsides.

Common uses: The stem bark is used to treat water swelling, swollen painful joints, and intestinal parasites.

Harvest time: The stem bark can be harvested at any time of the year.

Drying and processing: After stripping it from the stem, the bark is cut into pieces and dried in the sun.

Market potential: Low.

Note: This herb is known to be slightly toxic, especially in large doses.

Common Name: chocolate vine

Latin Name: *Akebia quinata* (Houtt.) Decne.
Chinese Name: mu tong 木通, wu ye mu tong 五葉木通 (stem), ba yue zha 八月炸 (fruit)

Part used: stem, fruit

Description: This deciduous vine has gray-brown bark and slender stems. The papery, dark green leaves are divided into five leaflets (occasionally three or seven) on long slender petioles (4.5–10 cm). Both the male and female flowers bloom from April to June, are slightly fragrant, and are a pale to dark purple color (sepals are occasionally white or light green). The fruit sets from June to August, is dark purple at maturity but has a white pulp surrounding shiny brown-black seeds. The fruit eventually dries and splits open to release seeds.

Biogeographical notes: This species is most common in disturbed areas, forest margins, along streams, and in meadows. It is most common from Massachusetts south to Virginia but can be found sporadically elsewhere in the Southeast.

Common uses: The stem is most commonly used for urinary difficulties, including water swelling, nervousness and anxiety with reddish urination, mouth ulcers, menstrual obstruction, insufficient lactation, and joint pain with swelling and inflammation. The fruit is used for fullness and pain in the chest and hypochondrium, painful or obstructed menstruation, feeling as if something is caught in the throat but there is no actual obstruction, and inhibited urination.

Harvest time: Stem can be harvested any time of the year; fruit is harvested when mature.

Drying and processing: Stem is sliced and dried in the sun. Fruit is cut in half and dried in the sun or by mechanical means (forced air).

Market potential: Medium. This is a common but not very popular herb in professional Chinese medicine. It can be found in a few commercial products but deserves wider usage.

Common Name: coco yam, wild taro
Latin Name: *Colocasia esculenta* (L.) Schott
Chinese Name: yu tou 芋頭

Part used: primarily rhizome; stem and flower also used

Description: This is a large plant with spongy leaf stems 30–80+ cm long with large green to dark green leaves with a red or purple spot where it attaches to the leaf stem. The large leaf is suggestive of an elephant's ear and this is sometimes used as another common name. It can flower any time from early spring through late autumn. The structure is known as a spathe, which is not a typical flower but like a jack-in-the-pulpit.

Biogeographical notes: This is a weedy plant found in streams, pond, canals, and other permanently wet areas along the coast of the Gulf of Mexico and has been found in several counties in California. It generally forms large colonies but cannot grow in areas with cold winters. It also is found in Central and northern South America, as well as some of the Caribbean islands.

Common uses: The rhizome is a common food in southern China and southeast Asia. When dried, the corm is used as a decoction to strengthen the digestive system, especially when there is poor appetite and feeling of exhaustion. It is also used to treat diabetes, scrofula, and externally (pounded into a paste) for scalds and burns.

The stems and flowers are also used in folk medicine in areas where the plants grow. The stems are boiled in decoction to treat rashes from nettle stings and skin inflammation from allergic reactions. They are also used to treat diarrhea and night sweating in children. The flowers are used to treat stomach pain with vomiting of blood.

Harvest time: The corm is harvested in autumn and either consumed fresh as a food or dried for use in medicine.

Harvest techniques: The corms are dug from the mud with an appropriate tool.

Drying and processing: All the roots and green stems are trimmed off the corm, then it is washed thoroughly and consumed fresh (after boiling) as a food or sliced and dried to use as a medicine. Some sources suggest using the fresh, uncooked roots to treat certain ailments. However, this should only be done by a professional practitioner since raw roots can cause throat irritation and burning.

Market potential: This is an established food and wild harvesting is likely viable for niche markets of specialty foods (i.e., restaurants, Asian markets, etc.). It is cultivated in some areas.

Notes: This is a very common food in the Pacific Islands and is considered sacred in Hawai'i. The corms are most commonly added to stews or steamed and either eaten whole or mashed and cooked in a variety of ways. This is a starch similar to potato. It can also be sliced and baked or fried similar to a potato chip.

Common Name: common reed
Latin Name: *Phragmites australis* (Cav.) Trin. ex Steud.
Chinese Name: lu gen 蘆根 (rhizome)

Part used: rhizome

Description: This robust reed springs from aggressive rhizomes with narrow stems (~6 mm wide) growing to 2+ m tall. It has long drooping leaves to 50 cm and 1–3 cm wide. This is a highly variable plant with several subspecies; it is generally taller in the South and shorter at high elevation or in northern Canada and Alaska.

Biogeographical notes: This plant grows throughout North America in moist places, particularly along riverbanks and lakes.

Common uses: The rhizome is very commonly used to treat fevers with thirst and irritation; stomach inflammation with vomiting and retching; lung inflammation with cough; lung abscesses with coughing of pus; scant, burning, and difficult urination; rashes of various etiologies; and food poisoning from fish.

Harvest time: Harvested from July to October.

Harvest techniques: The rhizome can form thick nets crisscrossing each other making harvest very difficult. Tools such as an adze, ax, or hatchet can be very useful. Roots and aboveground portions should be removed.

Market potential: Medium to high. Commonly used in Chinese medicine by practitioners and found in many traditional formulas.

Notes: This is an extremely safe botanical medicine with up to 1 kg of the fresh rhizome being used to treat lung abscesses. The rhizome is very lightweight and bulky, so large amounts of the fresh herb would need to be harvested to accumulate a viable amount to sell. Also, because it is so easy to grow, the imported product is very inexpensive, so it is not likely to offer great economic incentive to those interested in wild harvesting.

Common Name: fivestamen tamarisk

Latin Name: *Tamarix chinensis* Lour.
Chinese Name: cheng liu 檉柳

Part used: tender branches and leaves

Description: This shrub or small tree can grow to 8 m tall. It has foliage that is reminiscent of cedar, and is sometimes called saltcedar because of the leaf appearance and its ability to grow in salty conditions. The plant has a "wispy" appearance with reddish young stems and green leaves, topped by a pink inflorescence.

Biogeographical notes: Common from sea level to 2,500 m elevation in Western states from Wyoming, south into New Mexico and Arizona, and west into Nevada. Also found in California and sporadically in other Western states. It prefers riverways, lakeshores, and arroyos.

Common uses: This herb is used for various inflammatory skin conditions with papules and itching, as well as unresolved rashes. It is used for inflammatory joint conditions with swelling and pain; for this application the decoction can also be applied as a wash. Also used to treat common cold or influenza with fever and sweating.

Harvest time: Harvested from April to August.

Harvest techniques: The top 15–20 cm is harvested with hand pruners or another appropriate tool.

Drying and processing: Usually dried in the sun, although shade drying is recommended.

Market potential: Low to medium. Not used extensively in commercial products in the U.S. However, it has a reasonable body of literature and is worthy of further investigation. Related species have similar traditional uses.

Common Name: giant reed
Latin Name: *Arundo donax* L.
Chinese Name: lu zhu gen 蘆竹根

Part used: rhizome

Description: This is a strong reed growing 2–6 m tall from robust rhizomes. It is generally unbranched, appearing very similar to a bamboo, but like bamboo it can occasionally have slender branches growing from nodes along the stem. Its leaves are long (30–60 cm) and 2–5 cm wide, tapering to a slender tip.

Biogeographical notes: This common reed is found throughout the United States in moist areas such as meadows, streams, rivers, etc.

Common uses: The rhizome is used for a variety of inflammatory conditions including chronic inflammation with irritation and thirst, difficult urination with short, dark-colored voidings, toothache with gum inflammation, and coughing of blood. It is also used for physical exhaustion leading to low-grade fever and a sensation of heat deep in the body as if the bones are hot. The fresh rhizome is pounded into a paste to be applied externally for a variety of skin inflammations including nettles rash and other similar hot rashes.

Harvest time: Rhizomes are dug in September and October.

Market potential: Unknown. This is not a common herb used in professional Chinese medicine. However, it has very similar actions to another commonly used herb, *Phragmites communis* Trin.

Common Name: glossy privet
Latin Name: *Ligustrum lucidum* W. T. Aiton
Chinese Name: nü zhen zi 女貞子

Part used: ripe fruit

Description: This is a large shrub to small tree, growing to 25 m tall. The leaves are dark green and shiny. The flowers are small but a single tree will explode into a whitish array of tens of thousands of flowers. These flowers give way to small (7–10 mm) blue-black fruits that ripen to a red-black color, which are both medicine and thoroughly enjoyed by birds; the latter, unfortunately, helps to spread this weedy plant.

Biogeographical notes: The tree can be found sporadically in the Southeast, particularly in Florida and Louisiana; it is also found in a number of areas in California. It can grow in a wide variety of environmental conditions, which is part of the reason it has become such an invasive plant.

Common uses: This is an herb of significant importance, which is also backed by a strong body of modern research. Traditionally it is used to treat a variety of illnesses associated with a deficient state, including dizziness, soreness and achy pains of the lower back and knees, spermatorrhea, deafness (non-congenital), early graying of hair, body being overheated without infection (often in the late afternoon or early evening), and poor vision.

Harvest time: Mature fruits are harvested from plants that are at least four to five years old, usually in December.

Drying and processing: Fruits are dried in the sun.

Market potential: High. This is a major medicinal in Chinese medicine and used in many commercial preparations. Furthermore, there is strong evidence that this herb may have beneficial effects in a number of serious diseases including but not limited to hepatitis and kidney failure.

Common Name: greater celandine

Latin Name: *Chelidonium majus* L.
Chinese Name: bai qu cai 白屈菜

Part used: herb (aboveground)

Description: Plants can grow to 1 m but usually are around two-thirds that height. The blue-green leaves are characteristic of the poppy family and its bright yellow flower (up to 2 cm across) emerges in spring and summer and gives way to seed pods that yield small black seeds.

Biogeographical notes: Sporadic but common weed found in either moist or dry habitats, mostly in areas that have been disturbed by humans such as roadsides, railroad, fence lines, etc. It is found mostly in the northeastern and upper Midwestern U.S. and adjacent Canada but it can be expected elsewhere. The *Flora of North America North of Mexico* notes that it is difficult to distinguish from the native *Stylophorum diphyllum* (Michx.) Nutt. (celandine poppy or wood poppy) (Kiger, 1997). However, this native species is only found in the central Midwest and parts of Appalachia, north into

Michigan. The pedicel (flower stalk) of the native species is usually longer and the flower is generally larger with significantly longer styles (3–6 mm) than greater celandine (1 mm).

Common uses: Used to treat stomach and abdominal pain, intestinal inflammation with diarrhea, chronic cough, jaundice, water swelling, sores, and carbuncles. The plant can be used both internally or externally but should be used in low dosages. The fresh orange sap of the plant is known in Western herbal medicine as an external treatment for warts. The fresh or dried herb can be prepared as a decoction or tincture.

Harvest time: Greater celandine can be harvested at almost any time but is best from early flowering through the fruiting stage.

Harvest techniques: The top 15 cm or so of the plant is cut for drying. Due to the orange milky sap, gloves are recommended when harvesting.

Drying and processing: Many herbalists prefer to make a tincture with the fresh herb. The herb can also be dried in the shade with ample air-flow. The herb is dry when stems easily snap. Any large stems should be removed from the final dried material.

Market potential: While this herb is commonly used by many herbalists, due to its potential toxicity and low-dose nature, the amount used in commerce is relatively small.

Common Name: hairy willowherb
Latin Name: *Epilobium hirsutum* L.
Chinese Name: liu ye cai 柳葉菜

Part used: whole plant, including root

Description: A tall, slender, perennial plant with willow-shaped leaves, this wildlife attractor has reddish-pink flowers that are commonly visited by a number of different insect pollinators. The plant can grow to nearly 2 m and colonize large areas.

Biogeographical notes: This non-native *Epilobium* (also known as willow herb or fireweed) is found in the Pacific Northwest, the Midwest, and northeastern states of the United States, northward into Canada. It prefers full sun and tends to colonize places previously disturbed by humans.

Common uses: The whole herb is decocted for the treatment of hot diarrhea with a foul odor, feeling of fullness and pain in the abdomen with lack of appetite and slow bowel movements, toothache, irregular menstruation, stopped menstruation, vaginal discharge, traumatic injuries, hemorrhoids, and scalds and burns (externally). In addition, the leaves of this plant are drunk as a simple herbal tea in Russia.

Harvest time: Can be gathered any time of the year but most commonly collected in the late spring to early summer before flowering.

Harvest techniques: The whole plant can be pulled from the ground or cut at the base. While the root is generally included, leaving it behind is acceptable and saves significant energy when cleaning and processing (though if one is dealing with an unwanted invasive population, roots should always be pulled up).

Drying and processing: The plant should be dried in the shade, then cut into manageable pieces for preparing tea. The fresh herb can also be pounded and the juice used externally.

Market potential: Small market potential. However, the leaves make a pleasant tea with a mild anti-inflammatory action. It is safe to consume using this format and has potential to be a niche product.

Common Name: Japanese barberry
Latin Name: *Berberis thunbergii* DC.
Chinese Name: Riben xiao bo 日本小檗

Part used: rhizome bark and root

Description: This is a small, deciduous, compact, well-armed shrub that grows to approximately 1 m. Its dark red branches leaf out with abundant gray-green leaves 1–2 cm long. The thin spines grow 5–15 mm long; these, along with its compact growth pattern, make for a difficult to penetrate fence. The off-yellow flowers bloom in May and June, and give way to small (8 × 4 mm) shiny, red, fleshy berries. This species is distinguished from the common barberry (*Berberis vulgaris* L.) by three easily assessed characteristics: (1) this species has two to four flowers on an inflorescence while common barberry has 10–25, (2) the leaf edge of this species is entire while the common barberry is toothed, (3) this species' spines are single while the common barberry's spines are in groups of three.

Biogeographical notes: Japanese barberry is found in disturbed areas, floodplains, forests, meadows, and fields throughout most of New England, south through most of New Jersey and Pennsylvania, then sporadically west to Illinois and Wisconsin, but can be expected elsewhere. This is a highly invasive plant.

Common uses: This plant is used for a number of, mostly, acute inflammatory ailments including diarrhea, jaundice, stomach inflammation with pain, painful swollen red eyes, mouth sores, inflamed sore throat, acute eczema, and burns. It can be used either internally or externally.

Harvest time: Can be harvested any time but the best is early spring or late autumn.

Harvest techniques: Dig or pull the plants from the ground and cut the roots and rhizomes from the aboveground portions. Rhizome bark should be stripped while still fresh. The small spines are very sharp; one should be careful when harvesting.

Market potential: Good. This plant can be used as a substitute for some native plants such as Oregon grape root (*Mahonia* or *Berberis* spp.) and to a lesser extent goldenseal (*Hydrastis canadensis* L.).

Common Name: Japanese climbing fern
Latin Name: *Lygodium japonicum* (Thunb.) Sw.
Chinese Name: hai jin sha gen 海金沙根

Part used: herb

Description: This fern spreads with rhizomes that can go in any direction and from them sprout slender stems that can stretch out to set down new roots. The stems are densely covered with hairs and fronds emerge from the stems every 5–15 cm. The fronds are generally about 12 cm long and wide with five to seven lobes and are densely covered with short hairs. Spores are produced on the bottoms of short protrusions from the lobes.

Biogeographical notes: Common along the Gulf Coast and most of central and southern Louisiana, this plant can be found mostly in secondary vegetation in both shaded and partially sunny areas.

Common uses: This herb is used for a wide variety of inflammatory conditions such as urinary system ailments (e.g., urinary tract infections), liver inflammation, diarrhea, dysentery, common cold with fever, painful and swollen throat, mouth sores, red and swollen painful eyes, mumps, breast abscesses, skin rashes, scalding, red and inflamed skin conditions with itching, and joint pain. It is also used for water swelling, turbid urination, vaginal discharge, and bleeding due to traumatic injury.

Harvest time: The herb is harvested from July through October.

Market potential: Low. Not used in professional medicine in North America, however, could be useful at the community level.

Common Name: Japanese honeysuckle

Latin Name: *Lonicera japonica* Thunb.
Chinese Name: jin yin hua 金銀花 (bud), ren dong teng 忍冬籐 (stem)

Part used: flower bud, stem

Description: A common, semievergreen climber, Japanese honeysuckle can dominate areas with its slender, hairy, greenish young branches that mature into twisting, brown branches with peeling bark; they often become hollow with age. The leaves are thick, leathery, and hairy, often persisting during the winter, especially in the southern part of its range. The fragrant, tubular flowers bloom from April through June and are white, turning a yellowish-golden color with age. The small black fruits follow the flowers and may persist through the winter if they are not eaten by birds.

Biogeographical notes: Common from southern New England, south to Florida, and west Arkansas and Louisiana. It is also common in some parts of California and should be expected elsewhere. The plant can grow in a wide variety of ecosystems but is most common along forest edges, in meadows, in disturbed areas, and along streams.

Common uses: The flower bud is used for a wide range of infections and inflammatory diseases, including but not limited to the common cold with fever, influenza, lung infections, fever, purulent ulcers and sores, hot diarrhea with bleeding and pus, swollen and painful throat with the inability to swallow, and other similar ailments. This herb is used both internally and externally.

The stem is used for influenza with fever, swollen and red ulcers and sores, carbuncles, breast abscesses, inflammatory diarrhea with bleeding, and swollen, red, painful joints.

Harvest time: Flowers are harvested as buds, before they open. The stem is harvested during the dormant stage from autumn through winter.

Harvest techniques: Harvesting the flower buds can be tedious. They must be harvested by hand by simply pulling them off. When harvesting, flowers must not be packed together in large amounts or stored in plastic bags because they will begin to compost quickly and ruin the quality.

Drying and processing: Flowers should be dried as quickly as possible; forced air is the best method. Once completely dry, they must be stored in an air-tight bag in a cool area.

Market potential: High. The flower is a major medicinal plant used by many practitioners and product companies. The stem is used less but also has good market potential.

Common Name: Japanese knotweed

Latin Name: *Reynoutria japonica* Houtt.
Chinese Name: hu zhang 虎杖 (roots and rhizome, or plant)

Part used: roots and rhizome

Description: Naturalized throughout most of the U.S. and Canada, this plant is generally well-known but not well-liked. It is a perennial plant growing from a stout and aggressive rhizome that grows 1–2 m tall. The rhizomes send up numerous stems with deciduous ovate to elliptic shaped leaves. The small white to greenish flowers (usually all female) are borne in sprays of long slender spikes from June to September, then later give way to small (4–5 mm), shiny, black-brown seeds (technically achenes).

Biogeographical notes: While this plant can be found throughout most of the United States, it is uncommon in most of the Southeast and absent in Florida, Texas, New Mexico, Arizona, and Utah. The plant prefers moist areas such as streams, riverbanks, wet meadows, etc., but will grow nearly anywhere.

Common uses: This plant has been used in Chinese medicine for nearly 2,000 years for various ailments including stopped and painful menstruation, non-descending placenta, abdominal tumors, arthritic conditions, jaundice, putrid vaginal discharge, traumatic injury, swollen and painful sores with purulent pus, and externally for scalds and burns. This is a very well-researched medicinal plant with modern applications that include contributions to the treatment of high cholesterol, hepatitis, cancer, Lyme disease, and various inflammatory diseases.

This herb is contraindicated for use during pregnancy.

Harvest time: The root and rhizome are dug in the late spring through the summer until early autumn.

Harvest techniques: Digging the roots and rhizomes of this plant can be challenging due to the size and thick weave of rhizomes near the surface of the ground. Tools such as hatchets and axes have been employed to open an area to allow for further digging. The roots can penetrate deep into the soil.

Drying and processing: Root and rhizomes are washed clean and sliced for drying.

Market potential: High. Used by Chinese and Western herbalists alike. One of the main active constituents, resveratrol, is also found in many commercial herbal supplements.

Common Name: joint-head grass (see small carpetgrass)

Common Name: Korean lespedeza

Latin Name: *Kummerowia stipulacea* (Maxim.) Makino
Chinese Name: chang e ji yan cao 長萼雞眼草

Part used: whole herb (aboveground parts)

Description: This is a small many-branched annual plant of the pea family that can grow prostrate along the ground or erect, reaching upward to the sun. Its light purple banner/standard has small darker purple dots at the base and white wings and keel (the tip of the keel turns very dark purple). Flowers give way to small ovoid seed pods (legumes), which contain a single small black seed. The Chinese name "chang e" means "long calyx" suggesting that this species has a longer calyx than most related species.

Biogeographical notes: This plant is common in many parts of the eastern half of the United States. Introduced as a cover crop it can be invasive in some areas.

Common uses: Treats common cold, summertime "colds" with vomiting and diarrhea, jaundice, diarrhea, carbuncles and sores (both internally and externally), bloody urination, nose bleeds, traumatic injuries, and vaginal discharge.

Harvest time: The herb is collected in July and August and dried using standard techniques.

Harvest techniques: The entire herb is pulled from the ground, then the root is removed.

Market potential: Low.

Common Name: kudzu

Latin Name: *Pueraria montana* var. *lobata* (Willd.) Maesen & S. M. Almeida ex Sanjappa & Pradeep
Chinese Name: ge gen 葛根 (root tuber), ge hua 葛花 (flower)

Part used: tuberous root, flower, seed, stem, leaf

Description: Kudzu is a robust and aggressive climbing plant with hairy stems that can grow 8+ m long. The stems arise from tuberous roots, are woody at the base, and can grow up to 0.3 m a day. The toothless leaves have three lobes. The purplish-red flowers bloom from July through October and arise in long arrangements (racemes) from 15–30 cm long.

Biogeographical notes: Found as far north as southern New England, this plant is a well-known invasive species of the southeastern United States, particularly from Virginia south to South Carolina, Alabama, Louisiana, and Arkansas, but should be expected elsewhere. It can also be found sporadically in the Pacific Northwest. The plant can grow almost anywhere, is very drought tolerant, and is extremely aggressive.

Common uses: The root is primarily used for common colds (with fever) and influenza when the neck and shoulders are tight and sore; it can also be used to address thirst when a person is feverish owing to these ailments. It is commonly used for various types of skin rashes, particularly when there is incomplete expression of the rash. Finally, it can be very useful to help slow diarrhea and dysenteric disorders.

The flower is most famous for the treatment of excessive alcohol intake and modern research suggests it may be useful for helping people to stop drinking if they are attempting to go sober. It can also be used to treat those who drink excessively and suffer from vomiting blood or blood in the stool. It is sometimes used for anxious people who tend to run hot accompanied by thirst, headache, dizziness, fullness of the abdomen and chest, and vomiting and acid reflux.

The seed is used for those who have consumed too much alcohol and suffer from diarrhea.

The stem is used either dry or fresh, internally or externally, for swollen and infected sores, carbuncles, and similar infections.

The leaf can be used externally to stop bleeding due to traumatic injury.

Harvest time: The root is harvested in the winter (starting in November), preferably after there has been a frost and the leaves have died. Flowers are harvested after the first week in August before they have opened. Seeds are collected in the autumn when fully ripe. Stem and leaf can be harvested at any time of the year.

Harvest techniques: The tuberous roots of this plant can be very large and difficult to dig. Shovels, pick-axes, hatchets, and other similar tools may be necessary.

Drying and processing: Roots are cut into pieces and dried in the sun.

Market potential: Roots: high. This is a commonly used and popular herb. Flowers: medium. The flowers are not used extensively but there may be a small market.

Notes: The roots are also used to produce a thickening starch commonly used in culinary preparations in China, Japan, Korea, and other Asian countries.

Common Name: mimosa

Latin Name: *Albizia julibrissin* Durazz.

Chinese Name: he huan pi 合歡皮 (bark), he huan hua 合歡花 (flower)

Part used: bark and flower

Description: This is a common showy tree with an open crown that grows up to 16 m tall and found in a variety of human landscapes but has escaped into the wild. The early to mid-summer (May through July) blooms of brilliant pink flowers offer a spectacular, if brief, show.

Biogeographical notes: This is a common landscape tree but in many eastern and southeastern states (Virginia, South Carolina, Alabama, Louisiana, and Arkansas) it has become a serious invasive plant.

Common uses: This herb has a very pleasant, somewhat uplifting yet calming, effect on the mind that is non-habit-forming and treats a variety of ailments associated with anxiety, depression, run-of-the-mill stress, insomnia, etc. Both the bark and the flower are used for this purpose. The bark is also used for traumatic injuries (both internally and externally) and in the treatment of skin ulcerations with swelling. The flowers are also used to treat forgetfulness, fullness in the chest with indigestion and loss of appetite, and poor visual acuity or blurred vision.

Harvest time: Harvested from June to September.

Harvest techniques: Bark should be harvested from two- to three-year-old branches.

Drying and processing: Bark should be immediately removed from woody stems and cut into pieces for drying. Flowers are best fresh, either made into a tea or processed into a tincture for later use. The flowers can also be carefully dried for later use.

Market potential: Medium to high. This has become a somewhat popular herb in the United States, used by both Chinese herbalists and some Western herbalists. It is commonly used in commercial products and domestic sourcing is likely to be well-received by manufacturers and suppliers.

Common Name: orange eye butterflybush
Latin Name: *Buddleja davidii* Franch.
Chinese Name: da ye zui yu cao 大葉醉魚草

Part used: branches and leaves; root bark

Description: This bush is a deciduous to semievergreen shrub that grows up to 5 m tall. The opposite leaves are green on top and grayish, soft and woolly below. The small purplish flowers grow in large clusters and have a distinctive yellow to orange throat, which is how it gets its common name.

Biogeographical notes: Most commonly found in coastal areas in central and northern California, much of coastal Oregon, the Puget Sound area in Washington, southern Connecticut, Long Island, and adjoining areas. The plant prefers moist well-draining soil and full sun to partial shade.

Common uses: This herb is used to treat a wide variety of ailments including cough accompanied by chills and aversion to cold, joint pain, traumatic injury, ulcers and sores, and externally, as a wash, for vaginal itching and athlete's foot.

Harvest time: The branches and leaves are harvested from July through October and used either fresh or dried. The root bark is collected in the autumn and dried for later use.

Drying and processing: The herb is commonly used as an alcoholic extract. The Chinese name for the processed herb is, in fact, "wine medicine flower" (*jiu yao hua* 酒藥花).

Market potential: Low but deserves further investigation.

Note: This plant has some associated toxicity. Large doses should not be used and a trained professional should be consulted before using it internally.

Common Name: perilla mint
Latin Name: *Perilla frutescens* (L.) Britton
Chinese Name: zi su ye 紫蘇葉 (leaf), zi su zi 紫蘇子 (nutlets)

Part used: leaves and nutlets ("seeds")

Description: Perilla is an annual mint with large ovate to round, green to purple leaves (green leaves generally are purple underneath) that give off a unique minty aroma. The stems are angled and usually dark purple to burgundy, but sometimes green. The plant can grow as tall as 2 m under cultivation but is usually not much taller than 1 m in the wild. The small flowers are hairy and white to red to purplish. Each flower gives way to four small seeds (nutlets) that are brown with a black netted appearance.

Biogeographical notes: This plant has naturalized through much of the eastern United States with the exception of northern New England. It can grow in nearly any environment from full sun to shady areas. It prefers moist soil.

Common uses: The leaves are used as both food and medicine throughout Asia. In Korea, the leaves are commonly used as a sort of wrap for small pieces of barbequed beef. In Japan, they are commonly added to various sushi dishes and produce the traditional pinkish-purple color of pickled ginger. In China, the leaves are used for the common cold, indigestion, seafood poisoning with vomiting and diarrhea, and coughing. The seeds are used for the treatment of cough and asthma. The seed oil is also expressed and used as a specialty oil high in omega-3 fatty acids similar to flax seed oil. Also, in Korea, the toasted seed is pressed for the oil and used in some specialty dishes.

Harvest time: Leaves are collected any time before the plant begins to flower, seeds in the late summer or early autumn after they are ripe.

Harvest techniques: When harvesting seeds, the entire plant is cut and placed on a tarp to dry in the sun. Once completely dry, the plants are beat or pounded to free the seeds from the calyxes, then winnowed clean.

Drying and processing: Leaves are eaten fresh or can be dried for tea. Seeds are winnowed from the stiff calyxes and other plant matter and stored in a cool dry place.

Market potential: High. However, the plant is commonly cultivated and wild harvest is far more labor intensive, thus its cost is likely to be prohibitive under present conditions.

Common Name: porcelain berry

Latin Name: *Ampelopsis glandulosa* var. *brevipedunculata* (Maxim.) Momiy. (synonym: *Ampelopsis brevipedunculata* (Maxim.) Trautv.)
Chinese Name: Dongbei she pu tao 東北蛇葡萄

Part used: root bark

Description: This is a woody vine with tendrils that assist its climbing. The leaves are palmately divided, sometimes deeply so, and the leaf edges have sharp or rounded teeth. The small off-white to yellowish flowers bloom in July and August and give way to small, fleshy, green berries that turn pink, purple, or blue with an iridescent-like sheen.

Biogeographical notes: This plant is most commonly found from eastern Massachusetts, south through northern New Jersey and southeast Pennsylvania, but should be expected elsewhere in the northeastern part of the United States and adjoining Canada. The plant can live in a wide variety of ecosystems from disturbed areas to dense forests, but is most common near human-disturbed ecosystems, particularly near water such as streams, marshes, and ponds.

Common uses: The root bark of this plant is used most commonly for painful, swollen, and inflamed joints. It is also used as an anti-inflammatory for vomiting, diarrhea, mouth ulcers, injuries due to blunt force trauma and falls (including bleeding due to these injuries), swollen and inflamed sores, burns, and scalds. This herb can be used either internally or externally as a poultice (often used when fresh) or wash.

Harvest time: The roots are dug from September through November.

Drying and processing: Roots are first washed then the bark is removed and dried in the sun.

Market potential: Low. However, this herb has good potential at a community level for the treatment of many common ailments.

Common Name: puncturevine
Latin Name: *Tribulus terrestris* L.
Chinese Name: bai ji li (fruit) 白蒺藜

Part used: fruit

Description: This is an annual plant with small yellow flowers that spreads along the ground. The opposite leaves are evenly pinnately divided with 6–16 leaflets. The hard fruits start green and mature to a yellow-brown color, and sometimes break into up to five sections; they are heavily armed with spines, some of which are hardened and can be 4–6 mm long.

Biogeographical notes: Puncture vine can be found throughout the United States but is most common west of the Missouri River. The plant prefers dry, sandy, disturbed sites, roadsides, etc.

Common uses: Most commonly used for some types of headaches, dizziness, chest fullness with pain, breast tenderness (premenstrual), red eyes with corneal opacity, skin rashes (including scabies) with itching, vitiligo, ulcers, and scrofula.

Harvest time: Fruits are collected in August and September when they turn from green to yellow.

Harvest techniques: Fruits have sharp spiny protrusions; harvesting must be done with care. The plant also grows prostrate, meaning that it crawls along the ground, making harvesting somewhat challenging. Generally, the entire plant is harvested and dried in the sun, then the fruits are winnowed.

Drying and processing: After winnowing, the fruits (often at that point broken into separate segments, or mericarps) are dried completely in the sun, then tumbled to blunt the spiny protrusions.

Market potential: Medium to high. This is a commonly used herb in both professional medicine and consumer products.

Common Name: purple nutsedge
Latin Name: *Cyperus rotundus* L.
Chinese Name: xiang fu 香附

Part used: rhizome tuber

Description: This sedge is a perennial growing 15–90 cm tall with slender stolons and ellipsoidal tubers (the medicinal part). The leaves are basal, narrow (2–5 mm wide), and usually are about the same length or shorter than the main stem (culm), which, as for other sedges, is triangular. The flowering structure sits on top of the culm and is dark red to purple-brown.

Biogeographical notes: This plant is native throughout most of Asia, Africa, Australia, and parts of Europe. It is considered an invasive weed in the Americas and generally considered one of the worst weeds in the world. In the United States, it is found throughout Louisiana and neighboring southeastern Texas and southeastern Arkansas, much of Florida and South Carolina, as well as large parts of central and southern California and southwestern Arizona, but can be expected elsewhere, particularly in the Southeast.

Common uses: This herb is used extensively in Chinese medicine for a variety of gynecological illnesses including irregular menstruation, painful menstruation, excessive menstrual bleeding, and breast pain associated with menstruation, as well as postpartum illnesses. It is also used for digestive disturbances including abdominal pain, gastric reflux (sour regurgitation), upper abdominal fullness, and retching and dry heaves.

Harvest time: The tuber is dug in the autumn and spring, but the autumn material is preferred.

Harvest techniques: The tuber is small (2–3.5 cm long, 0.5–1 cm thick) and special equipment is used in commercial production to dig and process it.

Drying and processing: There are two basic methods of drying and processing this herb. The tuber can be rinsed of dirt and any buds cut off. Then, the tubers are roasted until the hairs and small rootlets are brittle and easily break off; care must be taken not to scorch the tuber. Then, by mechanical means, remove the hairs and rootlets. Traditionally this was done by putting the tubers between two reed mats and walking on them. However, today it is more commonly done by putting them in a tumbler until they are free of hairs. The other method, which is, perhaps, more common, is to boil the fresh tubers until the inside of the tuber is no longer white. The tubers are then set out in the sun to dry until they are 70%–80% dry, then placed in large woks and dry mix-fried until all the hairs are removed, as above, being careful not to burn the tubers. Once all the hairs are "burned" off, they are placed in the sun to complete the drying process.

Market potential: Very high. This is a major medicinal plant in Chinese medicine. The main issue is that because it is so easy to cultivate, it is relatively inexpensive in the market. However, if one wanted to, for example, reclaim an inundated field for planting other crops, the harvested and processed tubers could easily be sold.

Common Name: seaside rose

Latin Name: *Rosa rugosa* Thunb.
Chinese Name: mei gui hua 玫瑰花 (flower bud)

Part used: flower bud, fruit (hip)

Description: This is a distinctive rose with dense stem growth, each stem covered with an abundance of spines. The plant grows to about 1.5 m and can colonize an area via rhizomatous growth. The flowers are generally pink but can range from white to dark pink, and give way to an orange to red fruit (hip) that can swell to 2–2.5 cm in diameter.

Biogeographical notes: This shrub is primarily found on the coasts of New England but can be found in other locations, particularly near bodies of water. It prefers well-drained soils and can handle the salt-spray of the coastal areas.

Common uses: The flower bud is the primary part used in medicine. It is used for pain, discomfort, and fullness in the chest region and abdomen. It is also used for breast tenderness (particularly associated with menstruation), irregular menstruation, diarrhea, vaginal discharge, and various types of sores. The fruit is very high in vitamin C content and can be used to prepare preserves or other preparations as a nutritious supplement. However, while the flower bud is very common in practice, the hip is rarely used in Chinese medicine.

Harvest time: Flower buds are harvested when they have fully swelled but have not begun to open. Fruits are harvested when they are fully ripe.

Drying and processing: Buds should be dried by mechanical means (forced air) to ensure quick drying and preservation of the medicinal properties. Fruits can be processed fresh or dried and reserved for later use.

Market potential: High. This is a significant medicine and the buds and hips can also be used as an everyday tea.

Common Name: Siberian elm
Latin Name: *Ulmus pumila* L.
Chinese Name: yu bai pi 榆白皮 (inner bark)

Part used: inner bark, fresh juice of root bark, and fruit

Description: Siberian elm is a tree that can grow to about 25 m with grayish longitudinally ridged bark. The leaves are simple with serrated margins and grow up to 8 cm long and 3.5 cm wide. During the winter, its buds are brown to reddish brown. The seeds are encased in a light flimsy structure known as a samara, which first appears light green then turns to a whitish color before falling to the ground.

Biogeographical notes: The tree is resistant to Dutch elm disease and has been planted as a replacement for the American elm. However, it has become invasive and can be found in nearly every state in the continental United States. It is particularly common in Massachusetts, Kansas, Oklahoma, and New Mexico, but has a strong foothold in many other states. This tree can be found in a number of habitats, particularly disturbed areas, forest edges, meadows and fields, and shores of rivers or lakes.

Common uses: The inner bark of older branches is used to increase urine output and treat urinary tract infections; it is also used as a diuretic for water swelling. It can treat ulcers and carbuncles, especially on the back, as well as scrofula, and whitened bald patches on the head. Fresh juice of the root bark has sometimes been used as a substitute. The fruits are used, much less often than bark, to strengthen digestion and rid the body of excessive water, stop vaginal discharge, and kill parasites.

Harvest time: The best time to harvest it is the spring as the buds begin to swell and open. It can also be harvested in August and September.

Harvest techniques: Larger branches are cut from the tree (small young branches are discarded). The bark is stripped using a hooked knife specifically for bark harvesting, then the inner and outer bark are separated. The outer bark is discarded and the inner bark is dried in the sun for use.

Drying and processing: After drying, the inner bark is cut in sections for use as a decoction or other preparations.

Market potential: Low to medium. Not extensively used in the West but has strong regional usage across its native range.

Common Name: skunk vine

Latin Name: *Paederia foetida* L.
Chinese Name: ji shi teng 鷄屎藤

Part used: aboveground portion; root

Description: This is a vine growing to about 5 m long. The vines can be with or without hairs and the papery to somewhat leathery leaves generally arise alternately along the stem. The flowers range from a pale purple to a grayish white color. This plant has a very distinctive odor. In China, it is called "chicken shit vine," so if you think you found it, crush a leaf and you will be able to easily identify this plant!

Biogeographical notes: Although it can be found sporadically in the Southeast, it is primarily found in Florida; also found on four of the seven main islands of Hawaiʻi. Common in open forests, forest edges, and along streams.

Common uses: Used for flu-like symptoms in the summertime with fullness in the abdomen, lack of appetite, and fever. Also used for swollen painful joints, dysentery, jaundice, inflammation and swelling of the spleen and liver, intestinal abscesses, scalds, weeping skin rashes, inflammatory skin diseases, and snake bites.

Harvest time: Harvested in September and October.

Market potential: Low. Not commonly used.

Common Name: small carpetgrass, joint-head grass
Latin Name: *Arthraxon hispidus* (Thunb.) Makino
Chinese Name: jin cao 藎草

Part used: whole herb

Description: The appearance of this common annual grass is similar to a very small bamboo with clasping leaves, each with small hairs that protrude from the edge. The grass aggressively spreads, rooting at the nodes, but doesn't generally grow taller than 30–35 cm.

Biogeographical notes: The grass is most common in the Appalachian states, particularly Virginia and North Carolina, but can be found from New Jersey south to Louisiana. It prefers streamsides, moist meadows, and other moist areas including crop land.

Common uses: This plant is used for chronic cough, asthma, and inflammation of the throat, mouth, nose, lymph nodes, and mammary glands, as well as skin sores and ulcers (externally).

Harvest time: Harvested from July to September.

Harvest techniques: The entire plant is pulled from the ground and dried with the roots.

Market potential: Unknown. This is not a commonly used herb in professional circles but has good potential for self-care and community health.

Common Name: thorny olive
Latin Name: *Elaeagnus pungens* Thunb.
Chinese Name: hu tui zi 胡頹子 (plant), hu tui zi gen 胡頹子根 (root), hu tui zi ye 胡頹子葉 (leaf)

Part used: ripe fruit, root, leaf

Description: This densely branched evergreen shrub is armed and usually grows to 3–4 m tall. The leaf is between 5–10 cm long and is a scaly brownish below and waxy green above. The small, off-white-colored flowers are funnel-shaped and give way to a brownish to reddish fruit.

Biogeographical notes: The shrub can be found sporadically in disturbed areas, fields, forest edges, and roadsides in the Southeast from Virginia to Texas; also found in southeastern Massachusetts.

Common uses: The fruit is used to treat digestive problems when there is difficulty digesting food with fullness and lack of appetite. It is used as an astringent to stop diarrhea, stop excessive menstruation, and to treat bleeding hemorrhoids. The leaf is used to treat coughing with difficulty breathing similar to asthma, or chronic cough. The root is used to treat swollen painful joints, as well as pain due to traumatic injury. It is also used to stop bleeding, including vomiting blood and blood in the stool.

Harvest time: The ripe fruit is harvested from April through June. The root is harvested in the autumn and the leaf is harvested during the summer before flowering.

Harvest techniques: This is a well-armed plant; gloves and other protective clothing are recommended for harvesting.

Drying and processing: The whole fruit is dried in the sun. The root and leaves are dried in the shade.

Market potential: Low. Not used in professional medicine in the West but could serve local communities.

Common Name: torpedo grass
Latin Name: *Panicum repens* L.
Chinese Name: pu di shu 鋪地黍

Part used: root and rhizome

Description: This grass spreads primarily via pointed rhizomes, hence the name "torpedo grass." While the grass may only grow 30–125 cm tall, the rhizomes branch in different directions and can reach 6 m in length and penetrate deep into the soil. It often forms thick mats along waterways, including in the water, choking out native vegetation.

Biogeographical notes: Not cold tolerant, this grass is primarily found in Florida and along the Gulf Coast. It prefers areas where there is water, including streams, ponds, salt marshes, bogs, and sandy coastal habitats. However, it can also be found in drier habitats.

Common uses: Torpedo grass is used to treat high blood pressure (fresh root and rhizome is preferred); nose bleeds; purulent vaginal discharge (applied as a wash); urinary tract infections with painful, scant, and malodorous urination; sinusitis; parotitis; jaundice associated with hepatitis; venomous snake bites; and traumatic injury.

Harvest time: Harvested from July to October.

Market potential: Unknown to low. Little used in professional medicine but does have some minor research supporting traditional indications.

Common Name: tree of heaven
Latin Name: *Ailanthus altissima* (P. Mill.) Swingle
Chinese Name: chun pi 椿皮

Part used: root bark

Description: Found in nearly every state in the United States, this pesky deciduous tree grows up to 20 m tall. The tree has smooth bark and pithy branches. The leaves are oddly pinnate with 13–27 leaflets that are opposite or approaching opposite; they tend to be darker green toward the tips and have a somewhat foul odor when crushed. It has greenish-yellow flowers from April to May and produces pinkish-yellow fruits, which are a seed set in the middle of a wing, from August to October.

Biogeographical notes: This tree will grow almost anywhere up to ~2,500 m elevation but is most common on the eastern seaboard from Massachusetts to South Carolina, west to Illinois and Arkansas. It is also common in much of California and several other western states.

Common uses: Commonly used for acute diarrhea and dysentery caused by bacterial infections with or without bleeding. Also used for bloody urination, excessive menstruation, bleeding sores and ulcers, vaginal discharge associated with bacterial infections, parasites, and hemorrhoids. The seeds are less frequently used but can be employed for inflamed and painful joints, blood in the urine or stool, urinary tract infection, and vaginal discharge.

Harvest time: Harvested in April and May.

Harvest techniques: The roots are first dug by unearthing the roots around the tree and then cutting sections of the roots to remove for processing.

Drying and processing: The sections are peeled, removing the bark from the woody center, then the inner bark is scraped away before the pieces are laid, inner side up, to dry in the sun. Once the bark is mostly dry, it is cut into sections and then allowed to finish drying. The dried root bark is made into decoctions, pills, powders, and boiled into a very thick gruel to be applied externally.

Market potential: Low to medium. Although in relatively common use, it also is an inexpensive herb that is generally only used for somewhat serious medical conditions. However, in the community health care area for low income or underserved communities, this could be an important herb of which to be aware.

Common Name: trifoliate orange
Latin Name: *Citrus trifoliata* (L.) Raf.
Chinese Name: zhi shi 枳實, zhi qiao 枳殼

Part used: mature fruit

Description: This is a well-armed shrub or small tree that grows from 1 to 5 m tall. The young branches are green, flat, and ridged with stout spines that can grow to 4 cm long. Unlike other plants in the *Citrus* genus, leaves of this species are deciduous and the mature leaves are compound with three (or sometimes five) leaflets. The white flowers, which appear from May to June, give way to green fruits with a fuzzy outer surface reaching to 3–4.5 × 3.5–6 cm and maturing to a dark yellow in October and November; each fruit has between 20 to 50 seeds.

Biogeographical notes: This shrub or small tree can be found escaped sporadically in the Southeast, but has found a home in Louisiana (nearly every county) and much of Arkansas (most of the southern half of the state). It prefers well-drained acid soil conditions with ample sunshine.

Common uses: Fullness and distention in the chest and rib-sides; fullness and pain in the abdominal region; binding masses in the breasts; hernia with pain; pain in the testicles; traumatic injury; distention, fullness, and/or pain after eating or no desire to eat with these symptoms; constipation; and uterine prolapse.

Harvest time: July and August either when fruits are small and immature or later when they are fully mature. These are considered slightly different medicines.

Harvest techniques: Very carefully. This is a well-armed plant and the spines are stout and sharp. Leather gloves, or similar protection, are recommended.

Drying and processing: The fruit is sliced in half and dried in the sun.

Market potential: Medium to high. This herb is considered interchangeable with the more commonly used *Citrus* × *aurantium* L. However, the latter is official while this herb is the historically used material, now used much less (Pharmacopoeia Committee of China, 2020).

Common Name: white mulberry

Latin Name: *Morus alba* L.
Chinese Name: sang ye 桑葉 (leaf), sang zhi 桑枝 (twig), sang shen 桑葚 (fruit), sang bai pi 桑白皮 (root bark)

Part used: leaf, twigs, fruit, root bark

Description: Originally brought to the United States in an effort to start a silk industry (the leaves are used as food for silkworms), this shrub or tree (3–10 m tall) has become a serious invasive in many states in the east and Midwest. The bark is generally gray to dark gray but sometimes is a brownish color, and slightly furrowed. The leaves are a bright, sometimes even shiny, green. The flowers are insignificant (very small and yellowish green) but the fruit is about the size of a raspberry and generally white but can ripen to pink, red, or even purple.

Biogeographical notes: Common from southern New England, south to Virginia and then west through Ohio, Illinois, and Kansas. White mulberry is drought tolerant and can grow in many different soil and environmental conditions.

Common uses: The leaves are used for common cold, influenza, dry cough without phlegm, fever, dry throat with thirst, headache associated with fever and influenza, cough with painful chest, and swollen, red, and painful eyes associated with chronic illnesses attributed to the liver. The root bark is used to treat a variety of lung conditions associated with coughing and fullness of the chest with thick yellow phlegm associated with a lung infection or other inflammatory processes. It is also used to treat asthma when there is excessive phlegm impeding one's ability to breathe. This herb is also used to treat coughing of blood and difficult urination caused by a variety of inflammatory factors. The fruits are used as a general tonic for the blood, specifically for the liver and kidneys. The twigs are used to treat a variety of joint pains with swelling and inflammation.

Harvest time: Leaves are harvested October to December after the first frost. The root bark can be collected in both the spring (just as the tree's leaves are emerging) and autumn (after the leaves have fallen). Fruits are collected in the spring when they are ripe. Twigs can be collected at any time although the autumn twigs are preferred.

Harvest techniques: Because leaves are harvested after the first frost, a good shake of a smaller tree will usually cause significant numbers of leaves to fall and can then be easily collected from the ground. Root bark is often collected simply by digging the entire tree from the ground, or at least significant sections of roots, using pick-axes and shovels (heavy equipment can also be employed). Twigs, approximately 1 cm in diameter, are cut from the tree with an appropriate tool and sliced thin for drying.

Drying and processing: Leaves are dried using standard methods. The outer yellow layer of bark on the root bark must be removed before cutting into pieces for final drying. Fruits are very tender and require forced air drying to ensure rapid drying without mold or other potential problems. Twigs are simply cut to appropriate lengths and dried in the sun.

Market potential: High. This is a very commonly used medicinal plant in Chinese medicine. The fruit is also edible and delicious. The leaves and root bark are the most commonly used in medicine with the twigs being least commonly used.

Common Name: wild taro (see coco yam)

Common Name: winter creeper

Latin Name: *Euonymus fortunei* (Turcz.) Hand.-Mazz.
Chinese Name: fu fang teng 扶芳籐

Part used: young branches and leaves

Description: This plant grows to 10 m tall and can either be a shrub or grow similarly to a shrub but with aerial roots that allow it to climb. Sometimes it resembles a vine and doesn't take any particular form, climbing over rocks, trees, fences, etc. The plant is evergreen in the southern part of its range but leaves usually brown (although they may persist) during the winter in the northern part of its range. The abundant green leaves are simple with rounded toothed edges. The small flowers have variable colorings from greenish to pinkish to purplish and can also be light yellow to whitish. The small orange fruits appear from September to December and are favored by birds.

Biogeographical notes: Most common from southeastern Massachusetts to northern New Jersey, Kentucky, and can be expected anywhere east of the Mississippi River. Generally found in disturbed areas as well as forest edges and scrub land.

Common uses: This herb is most commonly used as a pain reliever for lower back and knee pain and soreness, swollen painful joints, broken bones, and pain with bleeding due to traumatic injury. Also used to stop bleeding for coughing of blood, vomiting of blood, and excessive menstrual bleeding. In gynecology, it is used for irregular menstruation and prolapsed uterus. Finally, it is used for chronic diarrhea and water swelling. This herb is commonly used as a tincture.

Harvest time: The branches and leaves are harvested from February through November.

Drying and processing: Cut into pieces and dry in the shade.

Market potential: Low. Not commonly used in professional medicine in North America, however, it has good potential for use for some purposes in a community setting.

Literature Cited

Kiger, R. W. 1997. "*Chelidonium*." *In* Flora of North America Editorial Committee (editors), *Flora of North America North of Mexico*, Vol. 3, Magnoliophyta: Magnoliidae and Hamamelidae. Oxford University Press, New York. http://www.efloras.org/florataxon.aspx?flora_id=1&taxon_id=106616

Nanjing University of Chinese Medicine (editor). 2009. *Grand Dictionary of Chinese Medicinals*, 2nd ed., 3 vols. Shanghai Century Publishing, Shanghai. [In Chinese.]

Pharmacopoeia Committee of China. 2020. *Pharmacopoeia of the People's Republic of China*. Chinese Medical Science Press, Beijing. [In Chinese.]

USDA, NRCS. 2022. The PLANTS Database. National Plant Data Team, Greensboro. https://plants.usda.gov/

Wu, Z. Y., P. H. Raven & D. Y. Hong (editors). 1994–2013. *Flora of China*, 25 vols. Science Press, Beijing; Missouri Botanical Garden Press, St. Louis. https://www.efloras.org/flora_page.aspx?flora_id=2

Ye, T. (editor). 2017–2022. *Flora of Chinese Medicinal Plants*, 13 vols. Beijing University Medical Publishing, Beijing. [In Chinese.]

CHAPTER 8

Some Invasive Species of Demonstrated Medicinal Value

Wendy L. Applequist

Numerous plant species have received some use as medicine. Plants that are widespread and common seem, probably for a variety of reasons, to be more likely to be used than related plants that are rarer. It is therefore unsurprising that many invasive species have recorded medicinal uses. These species might provide benefits through direct personal use or by being harvested to make botanical products for sale.

In the United States, under the 1994 Dietary Supplement Health and Education Act (DSHEA), herbal products are not regulated *as drugs*, though manufacturing is very heavily regulated. Products for internal use that cannot be sold as conventional foods, which require a conventional-food dose form (primarily tea) and a lack of health claims, are regulated as dietary supplements (other finished products for consumption, e.g., tinctures, capsules, pills, granules, or teas that make health claims). Supplements, unlike drugs and a few conventional foods, may not claim to diagnose, treat, mitigate, cure, or prevent disease, and manufacturers may not mention the existence of research into such benefits in their advertising.

In practice, though, supplements are frequently used to treat, mitigate, cure, or prevent disease. Some folkloric uses are probably not very effective, but most of the botanicals that are most popular in the U.S. have support from clinical trials for at least one popular use. Most of those trials, even for American botanicals, were done in Europe or Asia, where scientific culture was more willing to test benefits based on traditional knowledge. Though manufacturers can't mention such trials in their advertising, the herb-using public can find out about them, and positive clinical trial reports make knowledgeable people more willing to presume that a real benefit exists.

Therefore, invasive plants with high-quality evidence of benefit could be of most interest for possible use in domestic medicinal (or "supplement") products. If they are not already well-known, this demonstrated activity could make it easier to persuade practitioners and consumers to try them. In this chapter, I review a selection of invasive species with relatively strong scientific evidence of benefit for at least one medicinal use. This thematic focus is certainly not intended to imply that other traditional medicines are unworthy, as many are rationally used based upon traditional experience, or a combination of traditional knowledge and preclinical studies.

Initially this chapter was intended to survey all of a selected set of relatively significant invasive species that appeared in the second edition of *Herbs of Commerce* (McGuffin et al., 2000) and that had been the subject of at least one positive clinical trial that could be located via PubMed or Google Scholar. However, those limitations ultimately appeared to be arbitrary and undesirable. For some plants, only one or two very small positive trials, using a weak design or studying a locally available combination product, were located, and my subjective opinion was that such research added nothing substantive to traditional knowledge. For other plants, real or perceived toxicity would prevent their use in American commerce no matter how well bioactivity was proven. Contrarily, a few plants (primarily, purgative laxatives) that still have acknowledged pharmaceutical use have received no modern clinical testing, presumably because their effects were already so well understood for obvious reasons; these clearly merit inclusion. Furthermore, a couple of plants that were not listed in *Herbs of Commerce*, but known as conventional *foods*, merited consideration of their potential use for nutraceutical or "superfood" products. (It is a peculiarity of DSHEA that certain plants that legally should not be sold in small capsules may be sold in jars or bags to be eaten in quantity as foods.) Others are commonly used in topical products and have been studied for that purpose. Therefore, ultimately, inclusion of species is based on my subjective judgment of the value of the available published evidence that they have value.

The list of plants considered was derived from the Invasive Plant Atlas of the United States (https://www.invasiveplantatlas.org). This resource includes some plants that are better described as weedy rather than invasive (they are native to the area where their presence is complained of, or there is no evidence of ecological harm) or that have been found only in a tiny region of the U.S. Only species recorded from at least 100 counties, an arbitrarily selected limit, were considered widespread enough to be of interest. Of those, species with evidence that they are perceived as harmful in practice, including by listing on multiple states' invasive weed lists (rather than just one or two states that have extremely broad lists) or by having stated ecological harms such as formation of dense colonies or poisoning of livestock, were considered to be demonstrably invasive.

This resulted in a list of 231 species (or groups of species treated together). That list was further refined by limiting it to species that were listed in *Herbs of Commerce* (McGuffin et al., 2000), with the exception of two relatively familiar food plants: Russian olive, *Elaeagnus angustifolia* L., whose fruits are both edible and medicinal, and wild parsnip, *Pastinaca sativa* L. These criteria reduced the number of species of interest to 87. Partial searches on species excluded from the list suggested that very few had accessible clinical trial data anyway, probably because the U.S. market is sufficiently large and diverse that most species used widely enough to attract the interest of clinical researchers have been used here.

Clinical trials were sought by searching the Cochrane Library of clinical trials (https://www.cochranelibrary.com), PubMed (https://pubmed.ncbi.nlm.nih.gov), and Google Scholar (https://scholar.google.com) using the species name and, for the latter two, the word "trial." If this resulted in an inconvenient number of irrelevant hits, the phrases "clinical trial" or "randomized controlled trial" could be searched. Those for which relevant publications were found were flagged for further investigation of literature. Systematic reviews for species with multiple clinical trials were sometimes retrieved by these searches and were preferentially cited when available, as they frequently examined foreign literature that was not available through these searches. However, the purpose of this chapter was *not* to do a complete literature review of all, or indeed any, of the considered species, only to highlight species for which relevant literature favoring human use is available to Western stakeholders. If a large amount of supporting literature exists, only a few key papers are cited, especially review papers or meta-analyses if such were available.

Invasive species that appear to have the most suggestive clinical evidence of potential medicinal value include the following 25 species, presented in alphabetical order by scientific name.

Arctium minus (Hill) Bernh. (lesser burdock): No clinical trials were located for this species. However, searching PubMed retrieved several clinical trials relating to greater burdock, *A. lappa* L., reporting beneficial effects on metabolic syndrome (Ha et al., 2021), pain and biomarkers of inflammation, cholesterol levels, and blood pressure in people with arthritis (Alipoor et al., 2014; Maghsoumi-Norouzabad et al., 2016, 2019), and ulcer and *Helicobacter pylori* infection (Wu et al., 2010, a very small study; and in a multiherb product, Yen et al., 2018). The roots of both are eaten (that of *A. lappa* is a valued vegetable, called *gobo*, in Japan) and some consider that their medicinal uses are largely similar. Though the existing clinical trial data for *A. lappa* are not definitive, they might suggest that greater interest in both burdock species would be warranted.

Artemisia absinthium L. (wormwood): As the common name suggests, this plant is traditionally used as a vermifuge; that activity has been confirmed in animal studies but has not been pursued further, possibly because Western medicine now prefers pharmaceutical vermifuges. Wormwood is also traditionally used to make the potent liquor absinthe, controversial because its thujone content was believed to cause mental problems. Thujone is toxic in excess (as is wormwood oil), but it's been noted that the symptoms of "absinthism" are indistinguishable from those of alcoholism, suggesting that thujone was not really responsible for the harms of overindulgence in absinthe (e.g., Padosch et al., 2006). Small clinical trials report that powdered encapsulated wormwood can induce remission of symptoms and aid in discontinuation of steroids in Crohn's disease (Omer et al., 2007; Krebs et al., 2010). Wormwood is used in Unani medicine for hepatitis B, and small trials (not obtainable by this reviewer) are said to have been published, but the quality of evidence appears to be very low to date. A controlled trial reported that topical 3% wormwood ointment could relieve symptoms of knee arthritis with efficacy comparable to piroxicam gel (Basiri et al., 2017). It would appear that general anti-inflammatory effects are present.

Artemisia vulgaris L. (mugwort): This species has several traditional uses in Western practice, but the only clinical trial located reported that it was useful as a topical product to treat vaginal *Candida* infection (Ebrahimzadeh Zgemi et al., 2019). In Asian medicine mugwort is used in moxibustion, the practice of burning small chunks of the compressed herb on or near selected acupuncture points. Chinese mugwort (*A. argyi* H. Lév. & Vaniot) might be the most traditional species, but *A. vulgaris* is also frequently reported in use. Moxibustion's mechanisms of action are not well understood, but clinical trials have confirmed it to be useful for several purposes. A somewhat dated review of systematic reviews by a group generally skeptical of moxibustion (M. S. Lee et al., 2010), including some reviews by the same research group, emphasizes that results are often not consistent among studies (also seen in trials of many Western interventions). However, multiple small trials reported benefits in ulcerative colitis, stroke rehabilitation, and for reduction of side effects of cancer treatment. A use of particular interest is to turn a breech fetus before delivery. This indication has been the subject of the greatest number of clinical trials in Western countries, most though not all favorable, including one (Cardini & Weixin, 1998) published in the highly regarded conventional medical journal *JAMA*.

Berberis vulgaris L. (barberry): This traditional botanical, whose root bark contains the antibacterial compound berberine, is illegal to cultivate in the U.S. Though less invasive than the related *B. thunbergii* (Japanese barberry), it is an alternate host for the wheat rust fungus, a potentially devastating agricultural pathogen, making its extirpation a matter of national economic security. Almost all of the scientific research is from Iran. Barberry has had reported benefit for a variety of therapeutic uses in controlled trials, including improving liver function in non-alcoholic fatty liver disease (Kashkooli et al., 2015), easing the symptoms of opioid withdrawal (Dabaghzadeh et al., 2021), and enhancing the effect of topical mitronidazole for bacterial vaginosis (Masoudi et al., 2016). A meta-analysis of five controlled trials (Hadi et al., 2019) found significant reductions in total and LDL cholesterol and blood triglycerides. Across seven small trials, modest or inconsistent effects on measure of blood sugar control, including insulin levels and HbA1c, were noted (Safari et al., 2020); however, some of those studies may have included healthy people, in whom strong effects should never have been expected. Water extract of the dried fruit, which is edible and used as flavoring, improves blood glucose, insulin, and lipid markers in patients with type 2 diabetes (Shidfar et al., 2012). Fruit extract also improves acne when taken orally (Fouladi, 2012).

Cichorium intybus L. (chicory): Chicory is well-known as a medicine and food: endive is a cultivated form of chicory, and the roots are used as a caffeine-free coffee substitute and flavoring. The roots contain inulin-type fructans, a type of soluble fiber also found in Jerusalem artichokes that causes flatulence in some people but is considered to have a variety of health benefits. Chicory has recently attracted significant research interest for its potential as a healthful supplement or food ingredient. In placebo-controlled trials, consumption of chicory fiber, including in products such as snack bars, increases beneficial gut flora (e.g., Marteau et al., 2011;

Reimer et al., 2020; Kiewiet et al., 2021) and reduces chronic constipation and softens stools (e.g., Marteau et al., 2011; Micka et al., 2016; Buddington et al., 2017). Healthy people have lower increases in blood sugar and insulin after eating products such as yogurt and jelly made with chicory inulin or oligofructans, compared to traditional products (Lightowler et al., 2018). In women with type 2 diabetes, consumption for two months reduces blood sugar, HbA1c, blood pressure, and some liver enzymes (Abbasalizad Farhangi et al., 2016).

Chicory "coffee" contains beneficial phenolic compounds (as does *actual* coffee). One short uncontrolled study reported that consumption reduced blood viscosity, and its authors speculated that chicory consumption could reduce risk of cardiovascular thrombosis, or abnormal blood clotting (Schumacher et al., 2011). Although evidence is limited, it is possible that chicory might have some of the same long-term benefits as more popular hot beverages.

Chicory seed extracts have been shown to reduce HbA1c in people with diabetes, the effect being clinically significant (reduction of 1.18% over 12 weeks; Chandra et al., 2020). One trial has reported that seed, alone or with turmeric, modestly reduces BMI in people with non-alcoholic fatty liver disease (Ghaffari et al., 2019).

Cinnamomum camphora (L.) J. Presl (camphor tree): This tropical tree is the richest source of camphor, which has uses as an incense, insect repellent, food flavoring, and medicine, especially in topical formulations and as a decongestant or cough suppressant. It is an ingredient in Vicks VapoRub®, Tiger Balm®, and similar products. In China, borneol extracted from *C. camphora* is used in "compound salvia pellet," whose main ingredients are *Salvia miltiorrhiza* Bunge (danshen) and *Panax notoginseng* (Burkill) F. H. Chen (sanqi ginseng). This product has been tested for its use against chronic angina in dozens of Chinese clinical trials, most individually small but together now including over 4,700 patients; two meta-analyses have found that the herbal product appeared to be both more effective and safer than nitrates (Wang et al., 2006; Wei et al., 2019).

The combination of camphor with hawthorn, a plant of demonstrated value for cardiac conditions, has also been tested in Europe with favorable results (e.g., Schmidt et al., 2000). Most trials are for orthostatic hypotension or other chronic low blood pressure; a meta-analysis of four small trials involving the product Korodin® reported significant benefits, though the ideal dose and duration of treatment are unknown (Csupor et al., 2019a). It should be noted that camphor taken orally in excess is toxic, which might make its use in American dietary supplements challenging at present. While clinical trials of topical OTC products containing camphor are limited, there have been reports of benefit for tension headaches (Schattner & Randerson, 1996).

Cyperus rotundus L. (purple nutsedge): Rhizomes of this species are widely used in Asian medicine. Most Asian clinical trials have involved combination products and details are usually inaccessible. In one reasonably sized trial (100 patients per group), a Thai six-species formula was shown to be non-inferior to the pharma drug diclofenac for the short-term treatment of knee osteoarthritis (Koonrungsesomboon et al., 2020). Both a combination product and a single-herb *C. rotundus* extract have been reported to increase weight loss, relative to placebo, in three-month trials in people with overweight or obesity (Salunke et al., 2019; Majeed et al., 2022). Small short trials have reported benefits for a five-herb formula called Davaie Loban in mild to moderate Alzheimer's (Tajadini et al., 2015). Though human evidence is weak to date, there are also interesting preclinical data from rodents; further studies would be of interest.

Single clinical trials have reported that the oil, used topically, may reduce the growth of underarm hair, underarm hyperpigmentation, and the side effects of laser hair removal (Mohammed, 2014, 2022a, 2022b); however, one would like to see this research replicated in larger studies and by a second group of independent researchers.

Elaeagnus angustifolia L. (Russian olive): The fruit, which is edible, has been reported to improve markers of cardiovascular health in postmenopausal women (Shabani et al., 2021) and women with arthritis and weight problems (Nikniaz et al., 2016). Russian olive extract may reduce symptoms of knee osteoarthritis comparably to ibuprofen (e.g., Panahi et al., 2016; Karimifar et al., 2017) and reduce markers of chronic inflammation (Nikniaz et al., 2014). Though the texture of the fruit is not particularly appealing, it can be used in processed products such as jellies, and might have some "superfood" potential. Flower extract may increase sexual desire in women (e.g., Zeinalzadeh et al., 2019).

Frangula alnus Mill. [*Rhamnus frangula* L.] (frangula, glossy buckthorn): Bark of this species is widely used as a laxative. It has not been subjected to clinical trial testing in the era of modern trial methodology, probably because its effects are so obvious that they were already considered adequately proven. (Anyone who doubts its effectiveness can rapidly test it for him- or herself!) The related species *Frangula purshiana* (DC.) A. Gray ex J. G. Cooper (cascara sagrada) is among very few botanicals still included on the Food and Drug Administration's OTC drugs list, which also includes psyllium and senna, two other laxatives.

Hedera helix L. (English ivy): Ivy's most common medicinal uses are for coughs as an expectorant, respiratory infections, childhood asthma, and other bronchial complaints. In 2011, a review of 10 human studies involving acute respiratory infections (Holzinger & Chenot, 2011) found that ivy or ivy/thyme products consistently showed benefit. However, most of the included studies were observational, not randomized controlled trials. Such studies can be far cheaper and include records from far more people (here, a total of 17,463 patients), but are also subject to confounding if people who receive a treatment differ in other ways from those who do not, which is often the case.

A later review by Reckhenrich et al. (2018) identified 19 trials examining use for a variety of respiratory conditions between 1993 and 2018, only eight randomized and half of those controlled by comparison to ivy cough drops. Between the weak methods of many studies, the fact that patients in different studies had different conditions, and the fact that cough is inherently difficult to measure numerically, these authors were unwilling to draw any conclusions about benefit. However, they agreed that the plant appeared safe. The most recent review (Barnes et al., 2020) that limited its scope to human studies for acute illnesses published after 2010 and for asthma after 2002 located 13 relevant human studies: five controlled clinical trials, two open-label trials, five observational studies, and one retrospective study. The studies included both single-herb and traditional combination products, and most included children or both children and adults. That review concluded that evidence supported the ability of English ivy to reduce the frequency and intensity of cough, convert a dry cough to a productive cough, and reduce sleep disturbance and need for antibiotic prescriptions. Again, no significant safety issues were found. Though English ivy has not been tested against COVID-19, Barnes et al. pointed out that its demonstrated benefit for respiratory infections in general "justifies further research to better understand its applicability" to that disease.

Hypericum perforatum L. (St. John's wort): This botanical with traditional wound-healing activities is now usually used for treatment of depression and, thanks to its acceptance by conventional practitioners in Europe, has been subject to substantial research. A rigorous Cochrane Collaboration review in 2008 (Linde et al., 2008) identified 18 placebo-controlled clinical trials and 17 blinded trials that compared St. John's wort to a pharmaceutical drug. Cochrane takes a conservative attitude, especially to complementary medicines, so it is noteworthy that they found adequate evidence to conclude that tested St. John's wort products "a) are superior to placebo in patients with major depression; b) are similarly effective as standard antidepressants; c) and have fewer side effects than standard antidepressants." Two additional meta-analyses have been published later, one involving fewer studies and neither more definitive. St. John's wort is not a problem species in the eastern U.S., but can grow out of control in the dry West and harm livestock, as it causes photosensitivity when eaten in large quantity. This effect is hardly ever seen in humans. However, medicinal use does reduce blood levels of many pharmaceutical drugs by improving liver enzyme activity, so that people taking such drugs, especially those with a narrow therapeutic index, should check with a knowledgeable health care provider before taking St. John's wort.

Ligustrum lucidum W. T. Aiton (glossy privet): Fruit of this species is called Nü Zhen Zi in Traditional Chinese Medicine (TCM). TCM virtually always uses multiple species in combination. Nü Zhen Zi is a component of several formulas that have been tested in clinical trials. Kuan-Sin-Yin decoction, of which *Ligustrum* is a minor component, is reported to modestly improve liver function in people with chronic viral hepatitis B and C and to decrease viral load in hepatitis C (C.-J. Lee et al., 2013; C.-Y. Liu et al., 2016). The Lingmao or Bushen Jianpi formula has been reported in good-sized long-term studies to be a useful, safe adjuvant to pharmaceutical treatment of hepatitis B (Zhu et al., 2013; Zhang et al., 2022). Erzhi Tiangui granules are reported to improve embryo quality in older women undergoing in vitro fertilization (Sun et al., 2021 being the only available literature in English). Six months of treatment with a seven-herb formula of which *Ligustrum* was the first listed ingredient improved quality of life and reduced recurrence for up to three years of follow-up in patients being treated with chemotherapy for stage III colon cancer (Jia et al., 2021). Another formula is reported to contribute to management of hypertension (Xu et al., 2013).

Lonicera japonica Thunb. (Japanese honeysuckle): Flowers, known in China as Jin Yin Hua, are combined with forsythia fruits to treat and prevent viral respiratory infections in Traditional Chinese Medicine; the combination is called Yinqiao powder. A review of 15 trials (Fan et al., 2021) found the combination to be beneficial for pneumonia in addition to Western medicine. However, the reviewers found evidence of potential publication bias (specifically, possible failure to publish negative trials, which makes a product look better in a meta-analysis) and recommended more trials be done. Honeysuckle flower is also used in more complicated formulas. It is among three ingredients of Jinqi Jiangtang tablets, which are used for diabetes; a Chinese meta-analysis of 10 rather small trials concluded that there were significant effects and no evident publication bias, but that more, higher-quality studies were needed (Deng et al., 2020). Yin Zhi Huang granules, made from four plants including honeysuckle flower, are used for liver problems. A recent meta-analysis of 19 trials found them to be of significant benefit, in addition to phototherapy, for neonatal jaundice (Feng et al., 2022).

Melilotus officinalis (L.) Lam. (yellow sweet clover): One study of yellow sweet clover reported it to reduce edema following plastic surgery (Xu et al., 2008). Several European studies have focused on products that combine yellow sweet clover with gotu kola (*Centella asiatica* (L.) Urb.) and isolated flavonoids or tocopherol. Small studies have reported one such product to benefit chronic venous insufficiency (Cataldi et al., 2001) and another to benefit diabetic macular edema, with benefits seen through a three-year follow-up period (Forte et al., 2011, 2013). However, more research, including larger long-term studies, should be done to confirm their effects. A few small studies of proprietary Iranian oral and topical products for diabetic neuropathy and foot ulcers or other wounds are inadequate to demonstrate efficacy or safety.

Mentha spicata L. (spearmint): Spearmint is traditionally used as a milder-tasting, but in some ways weaker substitute for peppermint (*M. ×piperita* L.). Spearmint tea is reported to have an antiandrogenic effect in women with polycystic ovary syndrome (PCOS) (e.g., Grant, 2010), and an herbal combination product improved blood lipid profiles in PCOS in a single-blind study (Ainehchi et al., 2020). In a 90-day trial in older people with age-associated memory impairment, spearmint extract moderately improved memory performance and improved sleep, mood, alertness, and energy (Herrlinger et al., 2018). In healthy, active people, the same extract improved measures of sustained alertness and reactive agility (Falcone et al., 2018, 2019). Capsules containing small doses of either peppermint or spearmint essential oil may reduce nausea and vomiting due to chemotherapy (Tayarani-Najaran et al., 2013). A placebo-controlled pilot study of a formula containing three herbs with gastrointestinal uses reported benefit for irritable bowel syndrome (Vejdani et al., 2006). A study of spearmint tea for the symptoms of knee osteoarthritis found that a spearmint tea high in rosmarinic acid provided significantly greater relief than the "placebo" of an ordinary commercial spearmint tea (which, however, had some apparent benefits; Connelly et al., 2014).

Morus alba L. (white mulberry): The leaves are used medicinally, as animal fodder, and to raise silkworms; the root bark and the edible, tasty fruits are sometimes also used (Chan et al., 2016). Leaves are traditionally used for diabetes (Lim et al., 2021), with many supporting animal studies. A meta-analysis of 13 human clinical trials concluded that white mulberry leaf reduces postprandial glucose levels (Phimarn et al., 2017); one of the mechanisms appears to be a reduction in the digestion and absorption of starch (Jósefczuk et al., 2017). Since that review was completed, additional small trials have reported positive results (Taghizadeh et al., 2022), and a different review and meta-analysis of 13 trials (Jeong et al., 2022) concluded that mulberry leaf "can control blood sugar level" but considered the data inadequate to establish that it was "an effective intervention." That review reported no significant effect on glycated hemoglobin (HbA1c) in people with diabetes or hyperglycemia. However, of the four studies that measured that value, two were tiny (17 or 23 people in both groups combined), and in the other two, placebo-group participants also had mean HbA1c at or below 5.8, suggesting a level of drug (or lifestyle) treatment that made further improvement nearly impossible. A study that was not included in that review, which used more patients (80) with higher HbA1c levels, did report a significant reduction (Sohail et al., 2020). Mulberry leaf is also commonly included in combination herbal products used for diabetes, one of which has impressive reported activity, though in a study with a rather weak design (Chatterjee & Fogel, 2018).

A product containing alkaloids from mulberry twigs was reported in a recent 16-week, 200-person trial (Qu et al., 2022) to reduce HbA1c by 0.71% relative to the placebo group. Percentages of people reaching HbA1c levels below 7.0% and 6.5% were also much higher in the treatment group than the placebo group (respectively, 46.8% vs. 21.6% and 29.9% vs. 10.8%). More gastrointestinal side effects were noted in the treatment group, but no serious safety risks. The same researchers also performed a 600-person 24-week trial (Qu et al., 2021) comparing the mulberry twig alkaloid product to acarbose, an antidiabetic drug popular in China. The two were identical in efficacy, with the botanical product providing nonsignificantly greater reductions in HbA1c and significantly fewer side effects.

Mulberry bark is one of the ingredients in the Chinese herbal formula Ding Chuan Tang, used for respiratory problems, especially wheezing and asthma; an English-language double-blind trial showing benefit in children was located (Chan et al., 2006). A small trial reported that a product combining mulberry root bark with heartwood of *Acacia catechu* (L. f.) Willd. (cutch tree) could reduce delayed-onset muscle soreness from athletic activity (Yimam et al., 2018).

Perilla frutescens (L.) Britton (perilla): This mint relative has edible leaves that are used as medicine, as are the seeds. Leaf powder was reported in a six-month controlled trial to reduce blood pressure, in people with elevated blood pressure, and oxidized LDL cholesterol (Hashimoto et al., 2020). An aqueous leaf extract may benefit nonspecific gastrointestinal complaints (Buchwald-Werner et al., 2014). Nutraceutical products based on perilla have been reported to benefit hay fever in children and adults (Tanaka et al., 2004; Marseglia et al., 2019). Fried perilla seed is another ingredient in the Chinese formula Ding Chuan Tang, also containing mulberry bark, used for asthma (Chan et al., 2006).

Pueraria montana var. lobata (Willd.) Maesen & S. M. Almeida ex Sanjappa & Pradeep (kudzu): Kudzu root has a variety of medicinal uses. The use that has received the most human study is to help people with alcohol use disorders to abstain or drink less. Several clinical trials have reported that kudzu root supplementation reduces drinking in heavy drinkers, including people who were not interested in seeking treatment to reduce consumption (Lukas et al., 2005, 2013; Penetar et al., 2015). Kudzu (called Gegen in China) is among the ingredients of the Gegen Qinlian herbal formula, which a meta-analysis of five controlled trials concluded can assist in blood sugar management in people with diabetes (Ryuk et al., 2017). Its traditional combination with Danshen (Chinese salvia, *Salvia miltiorrhiza* Bunge) has been reported to improve markers of vascular health in coronary patients over an extended period (Tam et al., 2009).

Kudzu is rich in isoflavones and is often used, like soybean, for menopausal symptoms, though almost all clinical trials have used other *Pueraria* species that are popular in Asia. A flower extract from *Pueraria montana* var. *thomsonii* (Benth.) M. R. Almeida, used in Japan, and mandarin orange peel was found to reduce hot flashes and improve markers of bone health in menopausal women over 12 weeks (Kim et al., 2020). Extract of the same flower, alone, reduced BMI and visceral fat in a 12-week placebo-controlled trial in which it was given as a food additive (Kamiya et al., 2012).

Kudzu is used in Asian herbal formulas to benefit people who have had ischemic stroke. One of the active ingredients in kudzu is puerarin, which is used in China to treat stroke; it is generally given by injection as a purified pharmaceutical drug, not as a traditional medicine. The Cochrane

Collaboration (B. Liu et al., 2016) reviewed 20 Chinese clinical trials, some giving the treatment up to 10 days after stroke; in a meta-analysis of 16 trials with a total of 1,305 participants, the proportion who did not have improved function during follow-up was reduced by more than half with treatment. However, the review concluded that larger trials, with longer-term follow-up, were needed.

Reynoutria japonica Houtt. (Japanese knotweed): This plant, whose resveratrol-rich rhizome is the primary source of resveratrol used in dietary supplements, has confusing scientific nomenclature. It has been placed in three different genera, so most literature has recognized it as *Fallopia japonica* (Houtt.) Ronse Decr. or as *Polygonum cuspidatum* Siebold & Zucc. A recent meta-analysis of eight controlled trials of Chinese herbal formulas that included this species reported significant effects in increasing recovery rate from acute respiratory infections and reducing symptoms (Wang et al., 2022). However, the authors noted that most such trials also gave patients in both groups antibiotics as part of usual care and suggested that more Chinese trials of herbal formulas without antibiotics should be done. Numerous clinical trials report that purified resveratrol has various health benefits, potentially protecting against multiple diseases. It is not possible to presume that a less purified extract of the plant richest in resveratrol will have the same predicted benefits, but it certainly has economic value for the production of concentrated supplements.

Rhamnus cathartica L. (buckthorn): As for *Frangula alnus*, it appears that this very widely used laxative has not been subjected to any recent double-blind clinical trials, but the effects can be known beyond any reasonable doubt from older medical practice and human experience. A related species, *F. purshiana* (DC.) A. Gray ex J. G. Cooper (cascara sagrada), is still included in the FDA's OTC drugs review.

Silybum marianum (L.) Gaertn. (milk thistle): Milk thistle seed is used for liver diseases, especially as a concentrated extract of several flavonolignans that are collectively referred to as silymarin. Clinical trials of silymarin are relatively numerous and vary in results. The now over 15-year-old Cochrane Collaboration review and meta-analysis (Rambaldi et al., 2007), based on 13 randomized trials of any milk thistle product, concluded that there was insufficient evidence of benefit for all chronic liver diseases combined. However, some measures of improvement relative to control groups were *clinically* significant but not *statistically* significant, suggesting not that benefit was absent but that the existing studies were too small to be definitive. For example, "Liver-related mortality was significantly reduced by milk thistle in all trials (RR [relative risk] 0.50, 95% CI [confidence interval] 0.29 to 0.88), but not in high-quality trials (RR 0.57, CI 0.28 to 1.19)." A point estimate of a 43% reduction in liver-related mortality in high-quality trials certainly merits further investigation!

Another systematic review and meta-analysis based on 19 trials of silymarin products (Saller et al., 2008) concluded that there was no evidence of benefit for viral hepatitis, but that in alcoholic liver disease, a significant reduction in risk of death was seen. The Cochrane researchers' decision to treat very disparate liver conditions as a group was a non-standard methodology that might have obscured evidence of benefit. More recent meta-analyses have always focused on trials of silymarin. Two recent meta-analyses of the use of silymarin in non-alcoholic fatty liver disease (NAFLD), each analyzing data from eight trials (Zhong et al., 2017; Kalopitas et al., 2021), concluded that markers of liver function were improved, though it must be said that the effect is not strong enough to be confirmed in every trial.

An important use of the isolated flavonolignan silibinin is in the treatment of *Amanita* (death cap) mushroom poisoning, which causes severe liver damage. In Europe, it has been accepted for decades based on extensive clinical data that silibinin (also called silybin), given intravenously to provide a higher dose and greater bioavailability, greatly reduces mortality and need for liver transplant in mushroom poisoning (e.g., Saller et al., 2001; Enjalbert et al., 2002; Mengs et al., 2012). The product, Legalon® SIL, an intravenous form of a legal dietary supplement, remains unavailable in the U.S. except through emergency FDA approval as an investigational drug.

Another traditional use of milk thistle is as a galactagogue, a substance that increases milk production, hence its common name. A placebo-controlled clinical trial found that silymarin increased milk production in lactating mothers (Di Pierro et al., 2008), something previously demonstrated in animal studies. A product including silymarin and galega (goat's rue, *Galega officinalis* L.), another traditional galactagogue, significantly increased milk production in mothers of preterm infants (Zecca et al., 2016); the likelihood of being able to nurse exclusively was increased for at least three months (Serrao et al., 2018).

Tribulus terrestris L. (tribulus, puncturevine): Tribulus is often used to improve sexual performance or satisfaction. Limited evidence from five small clinical trials indicates that tribulus improves sexual function in women (Martimbianco et al., 2020). It may also improve menopausal symptoms, at least in the short term (Fatima & Sultana, 2017). Evidence for effect on sexual function in men is inconsistent, but one good-sized study using a standardized Bulgarian product reported significant benefit (Kamenov et al., 2017); variations in product quality or dose may affect results. Most studies that have examined effects on sperm quality in men with idiopathic fertility issues have reported benefit (Sanagoo et al., 2019). Tribulus is often marketed as enhancing testosterone and athletic performance, but clinical studies seeking to confirm these uses mostly find little or no benefit (e.g., Rogerson et al., 2007). Tribulus has been reported in one three-month study to reduce blood sugar in women with type 2 diabetes (Samani et al., 2016).

Verbascum thapsus L. (common mullein): Among the more popular folk uses of mullein flower is its use in herbal oils or ear drops for children's ear pain from otitis media. Two clinical trials published in pediatric journals concluded that multicomponent herbal ("naturopathic") ear drops did reduce pain as well as conventional products, but that pain generally improves within a few days in any case so that no treatment is really needed (Sarrell et al., 2001, 2003). A small controlled trial (Taleb & Saeedi, 2021) recently reported that mullein cream may speed the healing of episiotomy wounds—a procedure, however, which should no longer be routinely inflicted in the first place.

Viburnum opulus L. (crampbark, European cranberrybush): This species has a variety of traditional uses, including for urinary problems, though its use for gynecological issues is probably more common. A recent clinical trial has reported that it can ease passage of small distal ureteral calculi (kidney stones in the ureter), with expulsion rates statistically equal to the pharmaceutical comparator tamsulosin, but with far lower rates of emergency hospital admission and need for extra analgesics (Gok et al., 2021).

Vitex agnus-castus L. (lilac chaste tree, chaste tree): Chaste tree fruit is used for women's hormonal issues. There have been many clinical trials of chaste tree for premenstrual syndrome (PMS) and premenstrual dysphoric disorder (PMDD); for an early review, see van Die et al. (2013). Recent systematic reviews differ in the conclusions that their authors felt should be drawn from these studies. One systematic review based on eight clinical trials concluded that chaste tree appeared to be a safe and effective treatment for PMS and PMDD (Cerqueira et al., 2017). Another review (Csupor et al., 2019b) rejected 18 of 21 located trials for having failed to follow the CONSORT recommendations about how to characterize and describe herbal products used in studies; however, the remaining three trials that they did review included 520 participants and showed a clinically meaningful benefit.

Literature Cited

Abbasalizad Farhangi, M., A. Zare Javid & P. Dehghan. 2016. "The effect of enriched chicory inulin on liver enzymes, calcium homeostasis and hematological parameters in patients with type 2 diabetes mellitus: A randomized placebo-controlled trial." *Primary Care Diabetes* 10: 265–271.

Ainehchi, N., A. Khaki, E. Ouladsahebmadarek, M. Hammadeh, L. Farzadi, A. Farshbaf-Khalili, S. Asnaashari, H. J. Khamnei, A. A. Khaki & M. Shokoohi. 2020. "The effect of clomiphene citrate, herbal mixture, and herbal mixture along with clomiphene citrate on clinical and para-clinical parameters in infertile women with polycystic ovary syndrome: A randomized controlled clinical trial." *Archives of Medical Science* 16: 1304–1318.

Alipoor, B., L. Maghsoumi-Norouzabad, R. Abed, M. A. Eteraf Oskouei, B. Eftekhar Sadat & M. Asghari Jafarabadi. 2014. "Effect of *Arctium lappa* L. (burdock) root tea on clinical signs and symptoms in patients with knee osteoarthritis." *Current Topics in Nutraceutical Research* 12: 149–154.

Barnes, L. A., M. Leach, D. Anheyer, D. Brown, J. Caré, R. Lauche, D. N. Medina, T.-A. Pinder, A. Bugarcic & A. Steel. 2020. "The effects of *Hedera helix* on viral respiratory infections in humans: A rapid review." *Advances in Integrative Medicine* 7: 222–226.

Basiri, Z., F. Zeraati, F. Esna-Ashari, F. Mohammadi, K. Razzaghi, M. Araghchian & S. Moradkhani. 2017. "Topical effects of *Artemisia absinthium* ointment and liniment in comparison with piroxicam gel in patients with knee joint osteoarthritis: A randomized double-blind controlled trial." *Iranian Journal of Medical Sciences* 42: 524–531.

Buchwald-Werner, S., H. Fujii, C. Reule & C. Schoen. 2014. "Perilla extract improves gastrointestinal discomfort in a randomized placebo controlled double blind human pilot study." *BMC Complementary and Alternative Medicine* 14: 173.

Buddington, R. K., C. Kapadia, F. Neumer & S. Thies. 2017. "Oligofructose provides laxation for irregularity associated with low fiber intake." *Nutrients* 9: 1372.

Cardini, F. & H. Weixin. 1998. "Moxibustion for correction of breech presentation: A randomized controlled trial." *JAMA* 280: 1580–1584.

Cataldi, A., V. Gasbarro, R. Viaggi, R. Soverini, E. Gresta & F. Mascoli. 2001. ["Effectiveness of the combination of alpha tocopherol, rutin, melilotus, and centella asiatica in the treatment of patients with chronic venous insufficiency."] *Minerva Cardioangiologica* 49: 159–163. [In Italian.]

Cerqueira, R. O., B. N. Frey, E. Leclerc & E. Brietzke. 2017. "*Vitex agnus castus* for premenstrual syndrome and premenstrual dysphoric disorder: A systematic review." *Archives of Women's Mental Health* 20: 713–719.

Chan, C.-K., M.-L. Kuo, J.-J. Shen, L.-C. See, H.-H. Chang & J.-L. Huang. 2006. "Ding Chuan Tang, a Chinese herb decoction, could improve airway hyper-responsiveness in stabilized asthmatic children: A randomized, double-blind clinical trial." *Pediatric Allergy and Immunology* 17: 316–322.

Chan, E. W.-C., P.-Y. Lye & S.-K. Wong. 2016. "Phytochemistry, pharmacology, and clinical trials of *Morus alba*." *Chinese Journal of Natural Medicines* 14: 17–30.

Chandra, K., V. Jain, A. Jabin, S. Dwivedi, S. Joshi, S. Ahmad & S. K. Jain. 2020. "Effect of *Cichorium intybus* seeds supplementation on the markers of glycemic control, oxidative stress, inflammation, and lipid profile in type 2 diabetes mellitus: A randomized, double-blind placebo study." *Phytotherapy Research* 34: 1609–1618.

Chatterjee, S. & D. Fogel. 2018. "Study of the effect of the herbal composition SR2004 on hemoglobin A1c, fasting blood glucose, and lipids in patients with type 2 diabetes mellitus." *Integrative Medicine Research* 7: 248–256.

Connelly, A. E., A. J. Tucker, H. Tulk, M. Catapang, L. Chapman, N. Sheikh, S. Yurchenko, et al. 2014. "High rosmarinic-acid spearmint tea in the management of knee osteoarthritis symptoms." *Journal of Medicinal Food* 17: 1361–1367.

Csupor, D., R. Viczián, T. Lantos, T. Kiss, P. Hegyi, J. Tenk, L. M. Czumbel, et al. 2019a. "The combination of hawthorn extract and camphor significantly increases blood pressure: A meta-analysis and systematic review." *Phytomedicine* 63: 152984.

Csupor, D., T. Lantos, P. Hegyi, R. Benkő, R. Viola, Z. Gyöngyi, P. Csécsei, et al. 2019b. "*Vitex agnus-castus* in premenstrual syndrome: A meta-analysis of double-blind randomised controlled trials." *Complementary Therapies in Medicine* 47: 102190.

Dabaghzadeh, F., F. Sharififar, A.-M. Ahmadzadeh & S. Karami-Mohajeri. 2021. "The effects of *Berberis vulgaris* L. root extract on the opiate withdrawal syndrome and psychological factors: A randomized double-blind clinical trial." *Journal of Basic and Clinical Physiology and Pharmacology*. Preprint: https://doi.org/10.1515/jbcpp-2020-0327

Deng, Z.-Y., M.-J. Wang, Y.-H. Fan & M. Liu. 2020. ["Meta-analysis of effect of Jinqi Jiangtang Tablets on treating insulin resistance in type 2 diabetes."] *Zhongguo Zhong Yao Za Zhi* 45: 188–195. [In Chinese.]

Di Pierro, F., A. Callegari, D. Carotenuto & M. M. Tapia. 2008. "Clinical efficacy, safety and tolerability of BIO-C (micronized Silymarin) as a galactagogue." *Acta Bio-medica* 79: 205–210.

Ebrahimzadeh Zgemi, S., N. Golmakani, M. J. Asili, A. Naseri, Z. Mohebbi Dehnavi, Z. Kamali & A. Saber Mohammad. 2019. "A comparative study on the effect of artemisia and clotrimazole vaginal cream on vaginal candida infection on non-pregnant women in fertile age in Mashhad in 2016–2017: A triple blind clinical trial." *Journal of Medicinal Plants* 18: 170–182.

Enjalbert, F., S. Rapior, J. Nouguier-Soulé, S. Guillon, N. Amouroux & C. Cabot. 2002. "Treatment of amatoxin poisoning: 20-year retrospective analysis." *Journal of Toxicology: Clinical Toxicology* 40: 715–757.

Falcone, P. H., A. C. Tribby, R. M. Vogel, J. M. Joy, J. R. Moon, C. A. Slayton, M. M. Henigman, et al. 2018. "Efficacy of a nootropic spearmint extract on reactive agility: A randomized, double-blind, placebo-controlled, parallel trial." *Journal of the International Society of Sports Nutrition* 15: 58.

Falcone, P. H., K. M. Nieman, A. C. Tribby, R. M. Vogel, J. M. Joy, J. R. Moon, C. A. Slayton, et al. 2019. "The attention-enhancing effects of spearmint extract supplementation in healthy men and women: A randomized, double-blind, placebo-controlled parallel trial." *Nutrition Research* 64: 24–38.

Fan, Y., W. Liu, R. Wang, S. Du, A. Wang, Q. Xie & R. Yang. 2021. "Efficacy and safety of yinqiao powder combined with western medicine in the treatment of pneumonia: A systematic review and meta-analysis." *Complementary Therapies in Clinical Practice* 42: 101297.

Fatima, L. & A. Sultana. 2017. "Efficacy of *Tribulus terrestris* L. (fruits) in menopausal transition symptoms: A randomized placebo controlled study." *Advances in Integrative Medicine* 4: 56–65.

Feng, Q., Z. Huang, L. Su, Y. Fan, Y. Guan & G. Zhang. 2022. "Therapeutic efficacy and safety of Yinzhihuang granules with phototherapy in neonatal pathologic jaundice: An updated systematic review and meta-analysis." *Phytomedicine* 100: 154051.

Forte, R., G. Cennamo, M. F. Finelli, P. Bonavolontà, G. de Crecchio & G. M. Greco. 2011. "Combination of flavonoids with *Centella asiatica* and *Melilotus* for diabetic cystoid macular edema without macular thickening." *Journal of Ocular Pharmacology and Therapeutics* 27: 109–113.

Forte, R., G. Cennamo, P. Bonavolontà, A. Pascotto, G. de Crecchio & G. Cennamo. 2013. "Long-term follow-up of oral administration of flavonoids, *Centella asiatica*, and *Melilotus*, for diabetic cystoid macular edema without macular thickening." *Journal of Ocular Pharmacology and Therapeutics* 29: 733–737.

Fouladi, R. F. 2012. "Aqueous extract of dried fruit of *Berberis vulgaris* L. in acne vulgaris, a clinical trial." *Journal of Dietary Supplements* 9: 253–261.

Ghaffari, A., M. Rafraf, R. Navekar, B. Sepehri, M. Asghari-Jafarabadi & S.-M. Ghavami. 2019. "Turmeric and chicory seed have beneficial effects on obesity markers and lipid profile in non-alcoholic fatty liver disease (NAFLD)." *International Journal for Vitamin and Nutrition Research* 89: 293–302.

Gok, B., Y. T. Atik, B. Uysal, E. Koc, S. Tastemur & H. I. Cimen. 2021. "Gilaburu extract (*Viburnum opulus* Linnaeus) is as effective as Tamsulosin in medical expulsive therapy of distal ureteral calculi." *International Journal of Clinical Practice* 75: e14950.

Grant, P. 2010. "Spearmint herbal tea has significant anti-androgen effects in polycystic ovarian syndrome. A randomized controlled trial." *Phytotherapy Research* 24: 186–188.

Ha, M.-S., J. S. Yook, M. Lee, K. Suwabe, W.-M. Jeong, J.-J. Kwak & H. Soya. 2021. "Exercise training and burdock root (*Arctium lappa* L.) extract independently improve abdominal obesity and sex hormones in elderly women with metabolic syndrome." *Scientific Reports* 11: 5175.

Hadi, A., A. Arab, E. Ghaedi, N. Rafie, M. Miraghajani & M. Kafeshani. 2019. "Barberry (*Berberis vulgaris* L.) is a safe approach for management of lipid parameters: A systematic review and meta-analysis of randomized controlled trials." *Complementary Therapies in Medicine* 43: 117–124.

Hashimoto, M., Y. Tanabe, S. Hossain, K. Matsuzaki, M. Ohno, S. Kato, M. Katakura & O. Shido. 2020. "Intake of alpha-linolenic acid-rich *Perilla frutescens* leaf powder decreases home blood pressure and serum oxidized low-density lipoprotein in Japanese adults." *Molecules* 25: 2099.

Herrlinger, K. A., K. M. Nieman, K. D. Sanoshy, B. A. Fonseca, J. A. Lasrado, A. L. Schild, K. C. Maki, K. A. Wesnes & M. A. Ceddia. 2018. "Spearmint extract improves working memory in men and women with age-associated memory impairment." *Journal of Alternative and Complementary Medicine* 24: 37–47.

Holzinger, F. & J.-F. Chenot. 2011. "Systematic review of clinical trials assessing the effectiveness of ivy leaf (Hedera helix) for acute upper respiratory tract infections." *Evidence-Based Complementary and Alternative Medicine* 2011: 382789.

Jeong, H. I., S. Jang & K. H. Kim. 2022. "*Morus alba* L. for blood sugar management: A systematic review and meta-analysis." *Evidence-Based Complementary and Alternative Medicine* 2022: 9282154.

Jia, R., N. Liu, G. Cai, Y. Zhang, H. Xiao, L. Zhou, Q. Ji, et al. 2021. "Effect of PRM1201 combined with adjuvant chemotherapy on preventing recurrence and metastasis of stage III colon cancer: A randomized, double-blind, placebo-controlled clinical trial." *Frontiers in Oncology* 11: 618793.

Jósefczuk, J., K. Malikowska, A. Glapa, B. Stawińska-Witoszyńska, J. K. Nowak, J. Bajerska, A. Lisowska & J. Walkowiak. 2017. "Mulberry leaf extract decreases digestion and absorption of starch in healthy subjects—A randomized, placebo-controlled, crossover study." *Advances in Medical Sciences* 62: 302–306.

Kalopitas, G., C. Antza, I. Doundoulakis, A. Siargkas, E. Kouroumalis, G. Germanidis, M. Samara & M. Chourdakis. 2021. "Impact of Silymarin in individuals with nonalcoholic fatty liver disease: A systematic review and meta-analysis." *Nutrition* 83: 111092.

Kamenov, Z., S. Fileva, K. Kalinov & E. A. Jannini. 2017. "Evaluation of the efficacy and safety of *Tribulus terrestris* in male sexual dysfunction—A prospective, randomized, double-blind, placebo-controlled clinical trial." *Maturitas* 99: 20–26.

Kamiya, T., A. Takano, Y. Matsuzuka, N. Kusaba, M. Ikeguchi, K. Takagaki & K. Kondo. 2012. "Consumption of *Pueraria* flower extract decreases body mass index via a decrease in the visceral fat area in obese humans." *Bioscience, Biotechnology, and Biochemistry* 76: 1511–1517.

Karimifar, M., R. Soltani, V. Hajhashemi & S. Sarrafchi. 2017. "Evaluation of the effect of *Elaeagnus angustifolia* alone and combined with *Boswellia thurifera* compared with ibuprofen in patients with knee osteoarthritis: A randomized double-blind controlled clinical trial." *Clinical Rheumatology* 36: 1849–1853.

Kashkooli, R. I., S. S. Najafi, F. Sharif, A. Hamedi, M. K. Hoseini Asl, M. N. Kalyani & M. Birjandi. 2015. "The effect of berberis vulgaris extract on transaminase activities in non-alcoholic fatty liver disease." *Hepatitis Monthly* 15: e25067.

Kiewiet, M. B. G., M. E. Elderman, S. El Aidy, J. G. M. Burgerhof, H. Visser, E. E. Vaughan, M. M. Faas & P. de Vos. 2021. "Flexibility of gut microbiota in ageing individuals during dietary fiber long-chain inulin intake." *Molecular Nutrition & Food Research* 65: e2000390.

Kim, J. E., H. Jeong, S. Hur, J. Lee & O. Kwon. 2020. "Efficacy and safety of kudzu flower–mandarin peel on hot flashes and bone markers in women during the menopausal transition: A randomized controlled trial." *Nutrients* 12: 3237.

Koonrungsesomboon, N., S. Nopnithipat, S. Teekachunhatean, N. Chiranthanut, C. Sangdee, S. Chansakaow, P. Tipduangta & N. Hanprasertpong. 2020. "Clinical efficacy and safety of Thai Herbal Formulation-6 in the treatment of symptomatic osteoarthritis of the knee: A randomized-controlled trial." *Evidence-Based Complementary and Alternative Medicine* 2020: 8817734.

Krebs, S., T. N. Omer & B. Omer. 2010. "Wormwood (*Artemisia absinthium*) inhibits tumour necrosis factor alpha and accelerates healing in patients with Crohn's disease—A controlled clinical trial." *Phytomedicine* 17: 305–309.

Lee, C.-J., C.-H. Cheng, Y.-H. Li, C.-Y. Liu & C.-H. Hsu. 2013. "A Chinese medicine, Kuan-Sin-Yin decoction, improves liver function in hepatitis B virus carriers: A randomized, controlled study." *Journal of Alternative and Complementary Medicine* 19: 964–969.

Lee, M. S., J. W. Kang & E. Ernst. 2010. "Does moxibustion work? An overview of systematic reviews." *BMC Research Notes* 3: 284.

Lightowler, H., S. Thondre, A. Holz & S. Theis. 2018. "Replacement of glycaemic carbohydrates by inulin-type fructans from chicory (oligofructose, inulin) reduces the postprandial blood glucose and insulin response to foods: report of two double-blind, randomized, controlled trials." *European Journal of Nutrition* 57: 1259–1268.

Lim, W. X. J., C. S. Gammon, P. von Hurst, L. Chepulis & R. A. Page. 2021. "A narrative review of human clinical trials on the impact of phenolic-rich plant extracts on prediabetes and its subgroups." *Nutrients* 13: 3733.

Linde, K., M. M. Berner & L. Kriston. 2008. "St. John's wort for major depression." *Cochrane Database of Systematic Reviews* 2008: CD000448.

Liu, B., Y. Tan, D. Wang & M. Lium. 2016. "Puerarin for ischaemic stroke." *Cochrane Database of Systematic Reviews* 2016: CD004955.

Liu, C.-Y., P.-H. Ko, H.-R. Yen, C.-H. Cheng, Y.-H. Li, Z.-H. Liao & S.-H. Hsu. 2016. "The Chinese medicine Kuan-Sin-Yin improves liver function in patients with chronic hepatitis C: A randomised and placebo-controlled trial." *Complementary Therapies in Medicine* 27: 114–122.

Lukas, S. E., D. Penetar, J. Berko, L. Vicens, C. Palmer, G. Mallya, E. A. Macklin & D. Y-W. Lee. 2005. "An extract of the Chinese herbal root kudzu reduces alcohol drinking by heavy drinkers in a naturalistic setting." *Alcoholism, Clinical and Experimental Research* 29: 756–762.

Lukas, S. E., D. Penetar, Z. Su, T. Geaghan, M. Maywalt, M. Tracy, J. Rodolico, C. Palmer, Z. Ma & D. Y-W. Lee. 2013. "A standardized kudzu extract (NPI-031) reduces alcohol consumption in nontreatment-seeking male heavy drinkers." *Psychopharmacology* (Berl) 226: 65–73.

Maghsoumi-Norouzabad, L., B. Alipoor, R. Abed, B. Eftekhar Sadat, M. Mesgari-Abbasi & M. Asghari Jafarabadi. 2016. "Effects of *Arctium lappa* L. (burdock) root tea on inflammatory status and oxidative stress in patients with knee osteoarthritis." *International Journal of Rheumatic Diseases* 19: 255–261.

Maghsoumi-Norouzabad, L., F. Shishehbor, R. Abed, A. Z. Javid, B. Eftekhar-Sadat & B. Alipour. 2019. "Effect of *Arctium lappa linne* (burdock) root tea on lipid profile and blood pressure in patients with knee osteoarthritis." *Journal of Herbal Medicine* 17: 100266.

Majeed, M., K. Nagabhushanam, B. Bhat, M. Ansari, A. Pandey, S. Bani & L. Mundkur. 2022. "The anti-obesity potential of *Cyperus rotundus* extract containing piceatannol, scirpusin A and scirpusin B from rhizomes: Preclinical and clinical evaluations." *Diabetes, Metabolic Syndrome and Obesity* 15: 369–382.

Marseglia, G. L., A. Licari, G. Ciprandi & Italian Study Group on Pediatric Allergic Rhinoconjunctivitis. 2019. "A polycentric, randomized, double blind, parallel-group, placebo-controlled study on Lertal®, a multicomponent nutraceutical, as add-on treatment in children with allergic rhinoconjunctivitis: Phase I during active treatment." *Journal of Biological Regulators and Homeostatic Agents* 33: 617–622.

Marteau, P., H. Jacobs, M. Cazaubiel, C. Signoret, J.-M. Prevel & B. Housez. 2011. "Effects of chicory inulin in constipated elderly people: A double-blind controlled trial." *International Journal of Food Sciences and Nutrition* 62: 164–170.

Martimbianco, A. L. C., R. L. Pacheco, F. L. Vilarino, C. D. O. C. Latorraca, M. R. Torloni & R. Riera. 2020. "*Tribulus terrestris* for female sexual dysfunction: A systematic review." *Revista Brasileira de Ginecologia e Obstetricia* 42: 427–435.

Masoudi, M., M. R. Kopaei & S. Miraj. 2016. "Comparison between the efficacy of metronidazole vaginal gel and Berberis vulgaris (Berberis vulgaris) combined with metronidazole gel alone in the treatment of bacterial vaginosis." *Electronic Physician* 8: 2818–2827.

McGuffin, M., J. F. Kartesz, A. Y. Leung & A. O. Tucker. 2000. *Herbs of Commerce*, 2nd ed. American Herbal Products Association, Silver Spring, Maryland.

Mengs, U., R.-T. Pohl & T. Mitchell. 2012. "Legalon® SIL: The antidote of choice in patients with acute hepatotoxicity from amatoxin poisoning." *Current Pharmaceutical Biotechnology* 13: 1964–1970.

Micka, A., A. Siepelmeyer, A. Holz, S. Theis & C. Schön. 2016. "Effect of consumption of chicory inulin on bowel function in healthy subjects with constipation: A randomized, double-blind, placebo-controlled trial." *International Journal of Food Sciences and Nutrition* 68: 82–89.

Mohammed, G. F. 2014. "Topical *Cyperus rotundus* oil: A new therapeutic modality with comparable efficacy to Alexandrite laser photo-epilation." *Aesthetic Surgery Journal* 34: 298–305.

Mohammed, G. F. 2022a. "Topical *Cyperus rotundus* essential oil for treatment of axillary hyperpigmentation: A randomized, double-blind, active- and placebo-controlled study." *Clinical and Experimental Dermatology* 47: 534–541.

Mohammed, G. F. 2022b. "The effectiveness of *Cyperus rotundus* essential oil in reducing the side effects of laser hair removal." *Journal of Cosmetic Dermatology* 21: 1501–1505.

Nikniaz, Z., A. Ostadrahimi, R. Mahdavi, A. A. Ebrahimi & L. Nikniaz. 2014. "Effects of *Elaeagnus angustifolia* L. supplementation on serum levels of inflammatory cytokines and matrix metalloproteinases in females with knee osteoarthritis." *Complementary Therapies in Medicine* 22: 864–869.

Nikniaz, Z., R. Mahdavi, L. Nikniaz, A. Ebrahimi & A. Ostadrahimi. 2016. "Effects of *Elaeagnus angustifolia* L. on lipid profile and atherogenic indices in obese females: A randomized controlled clinical trial." *Journal of Dietary Supplements* 13: 595–606.

Omer, B., S. Krebs, H. Omer & T. O. Noor. 2007. "Steroid-sparing effect of wormwood (*Artemisia absinthium*) in Crohn's disease: A double-blind placebo-controlled study." *Phytomedicine* 14: 87–95.

Padosch, S. A., D. W. Lachenmeier & L. U. Kröner. 2006. "Absinthism: A fictitious 19th century syndrome with present impact." *Substance Abuse Treatment, Prevention, and Policy* 1: 14. https://doi.org/10.1186/1747-597X-1-14

Panahi, Y., G. H. Alishiri, N. Bayat, S. M. Hosseini & A. Sahebkar. 2016. "Efficacy of *Elaeagnus angustifolia* extract in the treatment of knee osteoarthritis: A randomized controlled trial." *EXCLI Journal* 15: 203–210.

Penetar, D. M., L. H. Toto, D. Y-W. Lee & S. E. Lukas. 2015. "A single dose of kudzu extract reduces alcohol consumption in a binge drinking paradigm." *Drug and Alcohol Dependence* 153: 194–200.

Phimarn, W., K. Wichaiyo, K. Silpsavikul, B. Sungthong & K. Saramunee. 2017. "A meta-analysis of efficacy of *Morus alba* Linn. to improve blood glucose and lipid profile." *European Journal of Nutrition* 56: 1509–1521.

Qu, L., X. Liang, G. Tian, G. Zhang, Q. Wu, X. Huang, Y. Cui, et al. 2021. "Efficacy and safety of mulberry twig alkaloids tablet for the treatment of type 2 diabetes: A multicenter, randomized, double-blind, double-dummy, and parallel controlled clinical trial." *Diabetes Care* 44: 1324–1333.

Qu, L., X.-C. Liang, G.-Q. Tian, G.-L. Zhang, Q.-L. Wu, X.-M. Huang, Y.-Z. Cui, et al. 2022. "Efficacy and safety of mulberry twig alkaloids tablet for treatment of type 2 diabetes: A randomized, double-blind, placebo-controlled multicenter clinical study." *Chinese Journal of Integrative Medicine* 28: 304–311.

Rambaldi, A., B. P. Jacobs & C. Gluud. 2007. "Milk thistle for alcoholic and/or hepatitis B or C virus liver diseases." *Cochrane Database of Systematic Reviews* 2007: CD003620.

Reckhenrich, A. K., A. Klüting & M. Veit. 2018. "Ivy leaf extracts for the treatment of respiratory tract diseases accompanied by cough: A systematic review of clinical trials." *HerbalGram* 117: 58–71.

Reimer, R. A., A. Soto-Vaca, A. C. Nicolucci, S. Mayengbam, H. Park, K. L. Madsen, R. Menon & E. E. Vaughan. 2020. "Effect of chicory inulin-type fructan-containing snack bars on the human gut microbiota in low dietary fiber consumers in a randomized crossover trial." *American Journal of Clinical Nutrition* 111: 1286–1296.

Rogerson, S., C. J. Riches, C. Jennings, R. P. Weatherby, R. A. Meir & S. M. Marshall-Gradisnik. 2007. "The effect of five weeks of *Tribulus terrestris* supplementation on muscle strength and body composition during preseason training in elite rugby league players." *Journal of Strength and Conditioning Research* 21: 348–353.

Ryuk, J. A., M. Lixia, S. Cao, B.-S. Ko & S. Park. 2017. "Efficacy and safety of Gegen Qinlian decoction for normalizing hyperglycemia in diabetic patients: A systematic review and meta-analysis of randomized clinical trials." *Complementary Therapies in Medicine* 33: 6–13.

Safari, Z., A. Farrokhzad, A. Ghavami, A. Fadel, A. Hadi, S. Rafiee, A. Mokari-Yamchi & G. Askari. 2020. "The effect of barberry (*Berberis vulgaris* L.) on glycemic indices: A systematic review and meta-analysis of randomized controlled trials." *Complementary Therapies in Medicine* 51: 102414.

Saller, R., R. Meier & R. Brignoli. 2001. "The use of silymarin in the treatment of liver diseases." *Drugs* 61: 2035–2063.

Saller, R., R. Brignoli, J. Melzer & R. Meier. 2008. "An updated systematic review with meta-analysis for the clinical evidence of silymarin." *Forschende Komplementarmedizin* 15: 9–20.

Salunke, M., J. Banjare & S. Bhalerao. 2019. "Effect of selected herbal formulas on anthropometry and body composition in overweight and obese individuals: A randomized, double-blind, placebo-controlled study." *Journal of Herbal Medicine* 17–18: 100298.

Samani, N. B., A. Jokar, M. Soveid, M. Heydari & S. H. Mosavat. 2016. "Efficacy of the hydroalcoholic extract of *Tribulus terrestris* on the serum glucose and lipid profile of women with diabetes mellitus: A double-blind randomized placebo-controlled clinical trial." *Journal of Evidence-Based Complementary & Alternative Medicine* 21: NP91–97.

Sanagoo, S., B. S. Oskouei, N. G. Abdollahi, H. Salehi-Pourmehr, N. Hazhir & A. Farshbaf-Khalili. 2019. "Effect of *Tribulus terrestris* L. on sperm parameters in men with idiopathic infertility: A systematic review." *Complementary Therapies in Medicine* 42: 95–103.

Sarrell, E. M., A. Mandelberg & H. A. Cohen. 2001. "Efficacy of naturopathic extracts in the management of ear pain associated with acute otitis media." *Archives of Pediatrics and Adolescent Medicine* 155: 796–799.

Sarrell, E. M., H. A. Cohen & E. Kahan. 2003. "Naturopathic treatment for ear pain in children." *Pediatrics* 111: e574–579.

Schattner, P. & D. Randerson. 1996. "Tiger Balm as a treatment of tension headache: A clinical trial in general practice." *Australian Family Physician* 25: 216, 218, 220 passim.

Schmidt, U., M. Albrecht & S. Schmidt. 2000. ["Effects of an herbal crataegus-camphor combination on the symptoms of cardiovascular diseases."] *Arzneimittelforschung* 50: 613–619. [In German.]

Schumacher, E., E. Vigh, V. Molnár, P. Kenyeres, G. Fehér, G. Késmárky, K. Tóth & J. Garai. 2011. "Thrombosis preventive potential of chicory coffee consumption: A clinical study." *Phytotherapy Research* 25: 744–748.

Serrao, F., M. Corsello, C. Romagnoli, V. D'Andrea & E. Zecca. 2018. "The long-term efficacy of a galactagogue containing silymarin-phosphatidylserine and galega on milk production of mothers of preterm infants." *Breastfeeding Medicine* 13: 67–69.

Shabani, M., A. Rezaei, B. Badehnoosh, M. Qorbani, M. Yaseri, R. Ramezani & F. Emaminia. 2021. "The effects of *Elaeagnus angustifolia* L. on lipid and glycaemic profiles and cardiovascular function in menopausal women: A double-blind, randomized, placebo-controlled study." *International Journal of Clinical Practice* 75: e13812.

Shidfar, F., S. S. Ebrahimi, S. Hosseini, I. Heydari, S. Shidfar & G. Hajhassani. 2012. "The effects of *Berberis vulgaris* fruit extract on serum lipoproteins, apoB, apoA-I, homocysteine, glycemic control and total antioxidant capacity in type 2 diabetic patients." *Iranian Journal of Pharmaceutical Research* 11: 643–652.

Sohail, Z., N. Bhatty, S. Naz, A. Iram & S. A. Jafri. 2020. "Effect of *Morus alba* (white mulberry) leaf on HbA1c of patients with type II diabetes mellitus." *Malaysian Journal of Nutrition* 26: 077–084.

Sun, J., J.-Y. Song, Y. Dong, S. Xiang & Q. Guo. 2021. "Erzhi Tiangui granules improve in vitro fertilization outcomes in infertile women with advanced age." *Evidence-Based Complementary and Alternative Medicine* 2021: 9951491.

Taghizadeh, M., A. M. Zadeh, Z. Asemi, A. H. Farrokhnezhad, M. R. Memarzadeh, Z. Banikazemi, M. Shariat & R. Shafabakhsh. 2022. "Morus Alba leaf extract affects metabolic profiles, biomarkers inflammation and oxidative stress in patients with type 2 diabetes mellitus: A double-blind clinical trial." *Clinical Nutrition ESPEN* 49: 68–73.

Tajadini, H., R. Saifadini, R. Choopani, M. Mehrabani, M. Kamalinejad & A. A. Haghdoost. 2015. "Herbal medicine Davaie Loban in mild to moderate Alzheimer's disease: A 12-week randomized double-blind placebo-controlled clinical trial." *Complementary Therapies in Medicine* 23: 767–772.

Taleb, S. & M. Saeedi. 2021. "The effect of the *Verbascum Thapsus* on episiotomy wound healing in nulliparous women: A randomized controlled trial." *BMC Complementary Medicine and Therapies* 21: 166.

Tam, W. Y., P. Chook, M. Qiao, L. T. Chan, T. Y. K. Chan, Y. K. Poon, K. P. Fung, P. C. Leung & K. S. Woo. 2009. "The efficacy and tolerability of adjunctive alternative herbal medicine (*Salvia miltiorrhiza* and *Pueraria lobata*) on vascular function and structure in coronary patients." *Journal of Alternative and Complementary Medicine* 15: 415–421.

Tanaka, H., N. Osakabe, C. Sanbongi, R. Yanagisawa, K. Inoue, A. Yasuda, M. Natsume, S. Baba, E. Ichiishi & T. Yoshikawa. 2004. "Extract of *Perilla frutescens* enriched for rosmarinic acid, a polyphenolic phytochemical, inhibits seasonal allergic rhinoconjunctivitis in humans." *Experimental Biology and Medicine* (Maywood) 229: 247–254.

Tayarani-Najaran, Z., E. Talasaz-Firoozi, R. Nasiri, N. Jalali & M. K. Hassanzadeh. 2013. "Antiemetic activity of volatile oil from *Mentha spicata* and *Mentha* ×*piperita* in chemotherapy-induced nausea and vomiting." *ecancermedicalscience* 7: 290.

van Die, M. D., H. G. Burger, H. J. Teede & K. M. Bone. 2013. "*Vitex agnus-castus* extracts for female reproductive disorders: A systematic review of clinical trials." *Planta Medica* 79: 562–575.

Vejdani, R., H. R. Mohaghegh Shalmani, M. Mir-Fattahi, F. Sajed-Nia, M. Abdollahi, M. R. Zali, A. H. Mohammad Alizadeh, A. Bahari & G. Amin. 2006. "The efficacy of an herbal medicine, Carmint, on the relief of abdominal pain and bloating in patients with irritable bowel syndrome: A pilot study." *Digestive Diseases and Sciences* 51: 1501–1507.

Wang, G., L. Wang, Z.-Y. Xiong, B. Mao & T.-Q. Li. 2006. "Compound salvia pellet, a traditional Chinese medicine, for the treatment of chronic stable angina pectoris compared with nitrates: A meta-analysis." *Medical Science Monitor* 12: SR1–7.

Wang, Z.-J., J. Trill, L.-L. Tan, W.-J. Chang, Y. Zhang, M. Willcox, R.-Y. Xia, et al. 2022. "*Reynoutria japonica* Houtt for acute respiratory tract infections in adults and children: A systematic review." *Frontiers in Pharmacology* 13: 787032.

Wei, H., Y. Wang, P. Jin, J. Li, S. Zhang, H. Jiang, Z. Wang & Y. Li. 2019. "Compound salvia pellet might be more effective and safer for chronic stable angina pectoris compared with nitrates. A systematic review and meta-analysis of randomized controlled trials." *Medicine* (Baltimore) 98: e14638.

Wu, Y.-C., L.-F. Lin, C.-S. Yeh, Y.-L. Lin, H.-J. Chang, S.-R. Lin, M.-Y. Chang, C.-P. Hsiao & S.-C. Lee. 2010. "Burdock essence promotes gastrointestinal mucosal repair in ulcer patients." *Fooyin Journal of Health Sciences* 2: 26–31.

Xu, F., W. Zeng, X. Mao & G.-K. Fan. 2008. "The efficacy of melilotus extract in the management of postoperative ecchymosis and edema after simultaneous rhinoplasty and blepharoplasty." *Aesthetic Plastic Surgery* 32: 599–603.

Xu, Y., H. Yan, M. J. Yao, J. Ma, J. M. Jia, F. X. Ruan, Z. C. Yao, et al. 2013. "Cardioankle vascular index evaluations revealed that cotreatment of ARB Antihypertension medication with traditional Chinese medicine improved arterial functionality." *Journal of Cardiovascular Pharmacology* 61: 355–360.

Yen, C.-H., H.-F. Chiu, S.-Y. Huang, Y.-Y. Lu, Y.-C. Han, Y.-C. Shen, K. Venkatakrishnan & C.-K. Wang. 2018. "Beneficial effect of Burdock complex on asymptomatic *Helicobacter pylori*-infected subjects: A randomized, double-blind placebo-controlled clinical trial." *Helicobacter* 23: e12469.

Yimam, M., S. M. Talbott, J. A. Talbott, L. Brownell & Q. Jia. 2018. "AmLexin, a Standardized brand of *Acacia catechu* and *Morus alba*, shows benefits to delayed onset muscle soreness in healthy runners." *Journal of Exercise Nutrition and Biochemistry* 22: 20–31.

Zecca, E., A. A. Zuppa, A. D'Antuono, E. Tiberi, L. Giordano, T. Pianini & C. Romagnoli. 2016. "Efficacy of a galactagogue containing silymarin-phosphatidylserine and galega in mothers of preterm infants: A randomized controlled trial." *European Journal of Clinical Nutrition* 70: 1151–1154.

Zeinalzadeh, S., A. A. Mohaghaghzadeh, F. Ahmadinezhad & M. Akbarzadeh. 2019. "Comparison of the effect of *Elaeagnus angustifolia* flower capsule and sildenafil citrate tablet female sexual interest/arousal disorder in clinical trial study." *Journal of Family Medicine and Primary Care* 8: 3614–3620.

Zhang, J.-H., X. Zhang, Z.-H. Zhou, X.-J. Zhu, C. Zheng, M. Li, S.-G. Jin, et al. 2022. "Bushen Jianpi formula combined with entecavir for the treatment of HBeAg-negative chronic hepatitis B: A multicenter, randomized, double-blind, placebo-controlled trial." *Evidence-Based Complementary and Alternative Medicine* 2022: 6097221.

Zhong, S., Y. Fan, Q. Yan, X. Fan, B. Wu, Y. Han, Y. Zhang, Y. Chen, H. Zhang & J. Niu. 2017. "The therapeutic effect of silymarin in the treatment of nonalcoholic fatty disease: A meta-analysis (PRISMA) of randomized controlled trials." *Medicine* (Baltimore) 96: e9061.

Zhu, X.-J., X.-H. Sun, Z.-H. Zhou, S.-Q. Liu, H. Lv, M. Li, L. Li & Y.-Q. Gao. 2013. "Lingmao formula combined with entecavir for HBeAg-positive chronic hepatitis B patients with mildly elevated alanine aminotransferase: A multicenter, randomized, double-blind, placebo-controlled trial." *Evidence-Based Complementary and Alternative Medicine* 2013: 620230.

CHAPTER 9

Invasive Common Reed as Valuable Bio-Resource: Lessons Learned from Europe

Franziska Eller

T HE COMMON REED (*Phragmites australis* (Cav.) Trin. ex Steud.) is a tall perennial wetland graminoid occurring globally in the form of genetically different lineages. The habitat of the endemic North American subspecies, *P. australis* subsp. *americanus* Saltonstall, P. M. Peterson & Soreng, has been invaded by common reed lineages originating from Europe and Africa, leading to an almost exponential range-expansion of the exotic lineages. Impacts of the invader, which is overtaking marsh communities and replacing native vegetation in freshwater and tidal wetlands, are reduced biodiversity and altered trophic support function across a broad consumer range. However, there are also benefits associated with the invading reed stands, such as increased protection from storm waves or soil erosion, water purification, or carbon sequestration.

Rather than trying to eradicate invasive common reed, which often is futile or even devastating to the ecosystem, I suggest to utilize the biomass of the species and control its productivity through harvesting, as is done in many European areas. Sustainable purposes of common reed biomass include feedstock for bioenergy like biomethane or biocombustion, raw material for paper production, or construction like roof thatching, agricultural application like fertilizer, or wastewater treatment. In Europe but also other areas, the concept of paludiculture, the cultivation of wetland plants on rewetted organic soils, is gaining momentum. The experience from this alternative form of biomass production, which is beneficial in particular due to its lowered greenhouse gas emissions and other ecosystem services, can be valuable for potential harvesting and processing of North American non-native common reed.

The Ecology of Common Reed— The Harms and Benefits of an Invader

There are few truly cosmopolitan plant species occurring on the globe. The common reed, *Phragmites australis*, is growing in wetlands on every continent except Antarctica. It is widespread especially in the temperate areas, less common in the tropics, and usually highly productive wherever its stands have established (Den Hartog et al., 1989). While its origin of formation remains controversial (Gorenflot et al., 1993; Connor et al., 1998; Clevering & Lissner, 1999), it has spread far beyond its natural geographical limits. The habitat of the endemic subspecies *P. australis* subsp. *americanus* has been invaded by common reed lineages originating from Europe and Africa, an introduction likely facilitated by ship traffic (Fig. 9-1; Saltonstall, 2002; Lambertini et al., 2012). Here, lineage is defined as the chloroplast DNA haplotype of the plant, which is inherited maternally and geographically structured in many plant species (Saltonstall, 2002). The best-studied introduced lineage, which was discovered first and originates in Europe and Asia, is called EU type, or haplotype M (Saltonstall, 2002; Lambertini et al., 2012). More exotic reed lineages have been discovered since, although the EU type seems to be the most abundant one so far (Guo et al., 2013). The almost exponential range-expansion of the exotic lineages into North America, primarily the East Coast, has gained massive interest from scientists, investigating possible reasons for the superiority of the invaders (Chambers et al., 1999; Meyerson et al., 2010; Cronin et al., 2015). Suggestions include clonal integration (Chambers et al., 2003), high salt tolerance (Burdick et al., 2001), great fitness with respect to global change factors and eutrophication (Mozdzer et al., 2010; Mozdzer & Megonigal, 2012), and low herbivory pressure (Cronin et al., 2015) as explanations for the plant's vigor,

FIGURE 9-1. Stand of invasive common reed in Heinz Nature Reserve, Darby Creek, PA, U.S.A. (Photo by author.)

resistance against adverse abiotic conditions, and rapid establishment. Attention has also been drawn to potential novel invasion into Asia from Europe (Lambertini et al., 2020; Liu et al., 2022).

The non-native European-Asian lineage (EU type or haplotype M) has had a tremendous impact on the native ecosystems, overtaking marsh communities and replacing native vegetation in freshwater and tidal wetlands (Chambers et al., 1999, 2003). The species is well-known as an ecosystem engineer, affecting tidal wetland hydrology by elevated sediment accretion rates (Rooth & Stevenson, 2000), and resulting in reduced biodiversity, due to the establishment of monocultural stands, which hinders the trophic support function of the invaded habitats across a broad consumer range (Meyerson et al., 2000). Despite the often reported negative impact on species diversity (but see Kiviat, 2013), invasive *Phragmites australis* still provides the ecosystem services well-known from the dominant marsh plant (Cronin et al., 2016; Kiviat, 2019). Thus, support of higher trophic levels, enhancement of water quality and sediment stabilization, as well as nutrient and carbon sequestration or microclimate improvement through establishment of accidental wetlands are beneficial provisions in *P. australis*–dominated wetlands, regardless of invasion status (Chambers et al., 1999; Kiviat, 2013). Compared to no marsh or other native species, the invasive common reed has a higher buffer capacity against wave and surge damage, and promotes sediment denitrification at higher rates in response to simulated storm additions of nitrate, and following hurricanes, emphasizing the plant's significant value for coastal protection (Sheng et al., 2021; Yacano et al., 2022).

The crucial function of the tall grass for soil stabilization became obvious in fall 2016 in the lower Mississippi River Delta, where an invasive scale insect (Roseau cane scale, *Nipponaclerda biwakoensis*) attacked the foundation species, causing it to collapse and leaving the wetlands in a state of erosion that accelerates the already rampant land loss (I. A. Knight et al., 2018). The symptoms of the infestation raised the concern of land managers, as reduced sedimentation and sediment infilling of navigation channels were likely results of the dieback. The reed community in the Mississippi River Delta consists of several non-native lineages, the most abundant being the so-called Delta type (Hauber et al., 2011; Lambertini et al., 2012). Experiments have shown that the Delta type had lower tolerance to scale infestation than the less abundant, co-occurring EU type (I. A. Knight et al., 2018; Cronin et al., 2020). Recommendations for marsh restoration in the Mississippi River delta have therefore included the non-native EU type that is less susceptible to fungal and scale infections, to avoid further deterioration of the affected coastal wetlands (Cronin et al., 2020; Bumby & Farrer, 2022).

Management efforts to fight the non-native common reeds in North America have included herbicide applications, mowing, burning, grazing, or covering of the area with plastic (Hazelton et al., 2014). Simple cutting is rather ineffective over the longer term, since cutting perennial rhizomatous grasses like common reed can stimulate growth and shoot density (Güsewell et al., 1998; Asaeda et al., 2006; Derr, 2008). Cutting is more efficient in combination with plastic and/or herbicide. Plastic sheeting, however, is labor-intensive and applicable only to small areas. Herbicides can indirectly affect the associated mollusk and algae communities, thereby shifting the trophic interactions in the ecosystem (Back et al., 2012). Biological control is controversial and not recommended. For one, the dieback of non-native common reed in the Mississippi River delta has shown that, once established, its ecosystem services exceed any potential beneficial effects of its removal (Schneider et al., 2022). Moreover, the possibility of a biological pest attack aimed toward invasive common reed on the native subspecies *Phragmites australis* subsp. *americanus* after field release cannot be excluded (Kiviat et al., 2019; but see also Blossey et al., 2020). Even if the mechanical removal of most of the invader is successful, the returning plant communities are not very similar to the native vegetation, owing to litter layers, soil moisture differences, and the season of removal management (Rohal et al., 2019).

The European Perspective—
Fighting the Dieback of a Native Resource

While management efforts in North America predominantly consist of eradication attempts, European freshwater wetlands have been facing a drastic reed dieback starting in the late '80s and '90s (Den Hartog et al., 1989). Researchers have suggested that a combination of changes in water level regime, eutrophication, grazing, mechanical damage, reduced sediment, or water body regulation were responsible causes (Kovács et al., 1989; Ostendorp, 1989; Armstrong et al., 1996a; Čížková et al., 1999; Clevering, 1999). The most pronounced dieback was observed in Eastern, Northern, and Central Europe: Germany, Switzerland, the Netherlands, Czech Republic, Poland, and to a minor degree Scandinavia (van der Putten, 1997). Recently, however, reedbed decline has also been reported from the Mediterranean area, probably associated with artificially altered hydrology (Gigante et al., 2014; Tóth, 2016). The effects of reed dieback on the ecosystems are dramatic. Lakeshore instability causes mudbanks leading to water pollution, bird population decline, and reduced littoral fish populations, all of which have direct or indirect economic consequences (Ostendorp, 1989; van der Putten, 1997). Waterway authorities at many places committed to restoring reedbeds, but often without success (Ostendorp, 1989). One example of a successful restoration method is the exclusion of aquatic herbivores such as geese, which can yield increased common reed growth (Bakker et al., 2018). Another example of an improved ecological state of reedbeds is harvesting or burning biomass. Avoiding litter accumulation by rotational cutting schemes has been shown to increase insect and other associated plant species biomass and biodiversity (Güsewell, 2003; Andersen et al., 2021a, 2021b). Once the vegetation has been restored, eutrophication will be mitigated through denitrification in the rhizosphere (Fujibayashi et al., 2020).

An important lesson from reed dieback, but also invasion ecology, is that no plant is an island—the inhabited ecosystem harbors conditions which are crucial for the survival and performance of the vegetation. Even though plants that become invaders do so because they supposedly share superior resource acquisition traits and high phenotypic plasticity (Richards et al., 2006; Davidson et al., 2011; Mozdzer & Megonigal, 2012), the invaded habitat typically has certain characteristics, too, which favor the invasion (Zedler & Kercher, 2004). Thus, anthropogenic disturbance and interference has been identified as an important factor of ecosystems that are more easily invaded than stable areas (Dukes & Mooney, 1999; Sciance et al., 2016). The rapid expansion of invasive *Phragmites* is favored in ecosystems that have been modified by humans with soil drainage and restricted tidal force. The consequent lowered porewater salinity and sulfide concentration and shortened submergence support the species' spread (Chambers et al., 2003). Other anthropogenic impacts of global change, such as elevated atmospheric CO_2 or eutrophication, will facilitate ecosystems to be invaded disproportionally by certain plant species or lineages, which thrive under these novel disturbances (Dukes & Mooney, 1999; Mozdzer & Zieman, 2010; Guo et al., 2013; Eller et al., 2014, 2017). Herbicide tolerance of invaders may increase due to interacting global change factors such as elevated atmospheric CO_2 (Hellmann et al., 2008). Moreover, their potential pests may be enabled to move accordingly and interact negatively with abiotic factors, which can have disastrous consequences for the ecosystem (I. A. Knight et al., 2020; Schneider et al., 2022).

From this perspective, there is little promise for the eradication of invasive species, and preventing them from entering a non-native habitat may be the best means to fight their negative impacts. The already established stands of invasive *Phragmites australis* can serve as renewable resources of manifold purposes.

Novel and Traditional Applications of Common Reed

Construction

Common reed is a traditional plant species used for thatching roofs or walls in predominantly Eastern and Northern Europe, Africa, and Asia (Fig. 9-2). The material is cheap, readily available, and abundant, and has been used by man since prehistoric times (Köbbing et al., 2013). A thatched roof built properly with good quality reed and proper maintenance has a longevity of up to 50 years or more (Wichmann & Köbbing, 2015). The harvest of common reed for any type of construction material takes place during winter, when the ground is frozen, to harvest the dry shoots (Wichmann & Köbbing, 2015). Winter harvesting allows non-wetland–adapted machinery to drive on the soil. Harvest by hand, using allen-scythes, or special equipment with pneumatic tires or tracks, and even mowing boats, are used in mild winters on wet, soft soil. Winter harvesting reduces conflicts with nature conservation issues, such as breeding birds. Moreover, the shoots are dry during winter, and their labile nutrient content is low, increasing their stability, which is desirable for thatch (Fig. 9-3a, b; Woehler-Geske et al., 2016).

FIGURE 9-2. Thatched fisherman house in Hamburg, Germany. (Photo by author.)

FIGURE 9-3. Harvested bundles of common reed at a shore of Fjand, Denmark (**a**, top) and nearby an overwintering barn, Nijmegen, the Netherlands (**b**, bottom). The bundles will be used for thatching. (Photos by author.)

FIGURE 9-4. Thatched birdwatcher's hut, Vejlerne Nature Reserve, Denmark. (Photo by author.)

New approaches for buildings integrating biodegradable resources have been proposed in recent years, to complement traditional techniques of weaving and bundling reeds to create entire buildings. Reed-based systems have been designed as a variety of structures, on idea basis and already implemented (Fig. 9-4; Bouza & Asut, 2020). With timber frames, compressed loose reed can be successfully used as horizontally laid wall insulation with short construction time (Miljan et al., 2013). In Denmark, the Wadden Sea Centre has been opened in 2017 as an iconic, completely thatched building that aligns with the flat surrounding marsh landscape. Dutch thatchers have pushed innovative building with common reed even more forward by covering roofs and house fronts with thatched walls, in an aesthetically pleasing fashion that ensures insulation from cold and fire protection (Jensen, 2020).

Agricultural Applications

Common reed can also be used as a fodder plant. As such, the plant is harvested during summer, when the fresh shoots have a high nutrient content and thereby high fodder quality (Köbbing et al., 2013). Silage produced by lactic fermentation can preserve the product. The pH of silage is low, which suppresses cellular respiration and prevents the degradation of proteins and vitamins. Although *Phragmites australis* can result in excellent silage quality with adequate nutrient content for ruminants such as goats, its lignin content is comparably high, and the silage can be partly rejected. Nonetheless, common reed can at least be partly incorporated into feedstuff of silage or roughage for ruminants as a cheap source of high nutritional quality (Baran et al., 2002; Monllor et al., 2020).

More often than harvesting reeds for fodder, the plant is grazed directly by buffalo, cattle, sheep, or horses, as a means of controlling its growth along the Baltic Sea coast. Grazing ensures the establishment of the typical, short saltmarsh grass vegetation that has a high potential for ground-nesting birds (Jeschke, 1987; Bernhardt & Koch, 2003; Sweers et al., 2013). In a North American freshwater marsh, grazing by goats significantly reduced the cover of the non-native reeds, and horses and cows ate the plant as well. The relatively low digestible dry matter content of reed, however, resulted in a lower energy content compared with other marsh species. Grazing mammals may therefore actively select other species, if given the choice. It is recommended to use high-intensity, short-duration, rotational grazing in small enclosures to focus on the dominant occurring fodder plant, non-native reed (Silliman et al., 2014).

Common reed has more application possibilities within agriculture, other than fodder. Livestock farmers in wetlands may benefit from the productive grass by using it as bedding in animal enclosures, at a competitive price and with similar results as cereal straw, under the same mulching conditions. Such "spontaneous vegetation" used in livestock farming assists climate-smart, sustainable agriculture at a low cost (Durant et al., 2020). Reed can be used as potential biogas feedstock (see next section), and the material left over from biogas production is a chemically beneficial by-product that can be used as organic manure, reducing the demand for synthetic fertilizers (Hansson & Fredriksson, 2004). Reed sludge has been shown to result in the lowest greenhouse gas emissions from soils fertilized with different digestates. Hence, its use as a substrate for biogas production and subsequent fertilizer, instead of inorganic fertilizers, may mitigate greenhouse gas fluxes from agricultural soils (Czubaszek et al., 2019).

Bioenergy
Reed Cultivation for Bioenergy
Generally, the use of perennial wetland plants as biogas and bio-combustion feedstock has been suggested to be highly sustainable, due to their high productivity, renewability, and cultivation methods supporting lowered greenhouse gas emissions (Wichtmann & Couwenberg, 2013; Jurasinski et al., 2020). Renewable biomass, produced non-competing with food crops, is

a critical aspect to contribute to solving the global energy and climate crisis, and to gradually replace fossil fuels. The carbon footprint of renewable feedstock is low, and even lower when produced in wet agriculture, since CO_2 release is suppressed under reducing soil conditions (Bartlett & Harriss, 1993; Brix et al., 2001). At an adequate, controlled water level, even CH_4 (methane) production is minimized, balancing carbon oxidation and reduction (Jurasinski et al., 2020).

In natural, *Phragmites*-dominated freshwater wetlands, CH_4 emissions can be high. Over the short term (decades), these ecosystems may therefore be regarded as a source for greenhouse gases, but as a sink if evaluated long-term (longer than a century), as the radiative forcing of CH_4 is less over the long term relative to CO_2, attenuating radiative forcing (Brix et al., 2001). Indeed, it was assessed that CH_4 radiative forcing does not undermine the climate change mitigation potential of peatlands supposed to be rewetted. It is not recommended to postpone rewetting, if long-term warming effects through continued CO_2 emissions are to be decreased (Günther et al., 2020). Salt marshes have generally considerably lower CH_4 emission rates than freshwater wetlands (Al-Haj & Fulweiler, 2020). In New England salt marshes invaded by common reed, methane release is mainly dependent on local abiotic conditions such as soil salinity, the methane concentration present in porewater, and solar radiation or temperature, and can vary greatly even short-term and on a small scale (Mueller et al., 2016; Martin & Moseman-Valtierra, 2015). Those local environmental and seasonal effects can have a greater impact on methane emissions than the origin of the vegetation (native or non-native; Mueller et al., 2016; Emery & Fulweiler, 2014). Nonetheless, the invasive lineage of common reed has in some studies been reported to have elevated methane emissions compared to other native vegetation.

Common reed is capable of altering its rhizosphere conditions with its characteristic deep root system (Brix, 1987; Moore et al., 2012), including decreased surface soil salinity (Windham & Lathrop, 1999). Moreover, reed is known to promote advective and diffusive fluxes of gases from soils to the atmosphere (Armstrong et al., 1996b; Brix et al., 1996). These ecosystem alterations could explain the observed elevated release of CH_4. Other studies in North American coastal marshes have shown that the presence of non-native reed had no direct effect on CH_4, and that the plant's removal even exacerbated greenhouse gas emissions (Martin & Moseman-Valtierra, 2017). By direct comparison in a controlled salt marsh study, the non-native reed lineage EU type had non-significantly higher methane emissions than the native lineage (p = 0.061; Mozdzer & Megonigal, 2013), indicating that favored CH_4 release from the soil is a feature of the species itself, rather than of the non-native lineage specifically.

Biomass quality and morphological characteristics of representative common reed populations have been investigated previously on a broad scale. Hence, desirable biomass characteristics for several bioenergy purposes can be found to initiate reed utilization as a nature-based solution for renewable energy production in areas where this is not yet practiced (Ren et al., 2019; Eller et al., 2020a, 2020b).

Biocombustion

High biomass productivity is crucial to obtain sufficient yield for biocombustion crops. Invasive common reed has large shoot diameters and therefore tall stems, as can be deduced from allometric relationships of morphological characteristics of grasses (Lu et al., 2016). This relationship translates into high biomass production. Biocombustion requires specific chemical conditions of the biomass, to avoid slagging, ash fouling, or corrosion (Jenkins et al., 1998; Lewandowski & Kauter, 2003). Reed pellets are suitable for use in furnaces built for wood pellets, with modification of some control parameters (Link et al., 2015). Also in terms of ash content, moisture, calorific value, and mechanical strength, reed pellets are very adequate and meet the quality requirements for granulated biofuel (Jasinskas et al., 2020). Common reed for combustion should be dry and is therefore harvested during winter or early spring, before the onset of green shoot growth. Reed can potentially be co-combusted with other biomass-based fuels (Kask et al., 2013), although common reed has been shown to have similar or even better combustibility when compared to other grasses such as *Miscanthus* (Wichtmann et al., 2013).

Biogas and Bioethanol

Biomethane produced from renewable biomass is a sustainable alternative to fossil fuels and can be used as vehicle fuel, as fuel in combined heat and power systems, or to produce electricity and heat, and the digestate can be utilized as fertilizer (Karlsson et al., 2017; Czubaszek et al., 2019). A C:N ratio between 20:1 and 30:1 in the plant material is optimal for biogas production. More nutrients are retained in green tissue, which is why a late summer harvest of biomass is appropriate for biogas production (Dioha et al., 2013; Giannini et al., 2016). Biomethane yield of *Phragmites australis* is comparable to the biofuel species *Arundo donax* or other wetland plants, but not as high as the yield of conventional biofuel crops such as maize or sunflower (Mayer et al., 2014; Lizasoain et al., 2016; Baute et al., 2018; Eller et al., 2020a). When using common reed for biogas, the harvest time will be of high importance. The tissue quality, like hemicellulose or protein concentration, varies by species in general, not merely within *Phragmites*, and it affects the decomposition and biogas production to a high degree (Baute et al., 2018). Utilization of common reed for bioethanol is less investigated, but the results are promising, if suitable pretreatments are applied. Several studies show that common reed can be considered as a candidate material for ethanol, both derived from its natural habitat but also cultivated as a dedicated energy crop (Szijártó et al., 2009; Cavalaglio et al., 2016; Tozluoglu, 2018).

Pulp and Paper Production

The presumable largest reed field of the world (over 1,000 km^2) is located in the Liaohe River Delta in China. The reed biomass there is harvested to produce pulp for paper production. Yields have been increasing since the field is managed by freshwater irrigation (Brix et al., 2014). Paper can be produced from wood or non-wood pulp. The main raw materials for wood pulp are hardwood and softwood chips, of which 2–2.5 tons are required to make 1 ton of pulp (Croon, 2014). For the production of 1 ton of pulp, around 2.4 tons of dry reed are required (Chivu, 1968). The high proportion of short fibers in reed is desirable for pulp production, and the material is harvested in winter to assure a low water content in the biomass. High quality paper can contain up to 30% reed, while paper consisting of up to 80% reed can be used as lower-quality wrapping material (Köbbing et al., 2013).

Wastewater Purification

A major application of the versatile grass is phytoremediation. Domestic, industrial, and agricultural wastewater, heavy metals, pharmaceuticals, organic aromatic compounds, pesticides, or biotoxins can be degraded or (at least partially) removed in constructed wetlands (Fig. 9-5; R. L. Knight et al., 1999; Borin et al., 2001; Calheiros et al., 2009; Vymazal, 2011; Milani & Toscano, 2013; Vymazal & Březinová, 2016; Bavithra et al., 2020). One of the first documented engineered treatment wetland systems was patented as early as 1901 by a U.S. citizen (Monjeau, 1901). Herein, rural wastewater purification based on natural wetland principles is described through percolation of the water through a vegetated layer followed by different sediments. However, the technology is documented to have been first used and spread in the 1950s in Germany, by Käthe Seidel (Seidel, 1953, 1961, 1964).

FIGURE 9-5. Delegation of researchers visiting treatment wetlands for rural wastewater treatment, Nijmegen, the Netherlands. Vegetation is *Iris pseudacoris* (center back) and *Phragmites australis* (right back). (Photo by author.)

Seidel carried out experiments using different macrophytes growing on highly permeable substrates in modulated basins to improve inefficient rural treatment systems. The wastewater was purified in consecutive stages, and resulted in a constructed wetland model later commonly known as "Root Zone Method." The system was planted with *Phragmites australis* (Seidel, 1965; Kickuth, 1980).

Wetland plants serve many important purposes in constructed wetlands: large debris is filtered by stems and roots; the sediment is oxygenated through aerenchyma, fueling aerobic microbial processes; the water current and wind velocity are reduced; light at the water-sediment surface is attenuated; nutrients are taken up and stored; sediment erosion is minimized; and often systems are designed to also be aesthetically pleasing (Brix, 1994, 1997). Vegetated treatment wetlands are efficient low-cost systems and often a preferred solution for pollutant removal in the tropics and subtropics, since the vegetation period is long and the climate beneficial for high biomass production. In temperate regions, constructed wetlands are for example used to treat food and wine processing industry, mining waters, dairy, pig, or fish farm effluents, and landfill leachate (Vymazal, 2011, 2013).

Phragmites australis is now by far the most commonly used species for constructed wetlands, in particular in horizontal subsurface flow systems. The species is tolerant to high organic and nutrient loadings and has large belowground organs even in partially anoxic conditions, as is required by plants used in constructed wetlands designed for wastewater treatment (Kvet et al., 1999). In tannery wastewater treatment in Spain, common reed has been shown to have a great nutrient absorption capacity, which helped to reduce the pollution load (García-Valero et al., 2020). The plant also had high removal of paracetamol from a constructed wetland in Italy, which could be attributed to degradation associated with root biofilms (Ranieri et al., 2011). In the Czech Republic, common reed was shown to be capable of removing trace metals from municipal sewage to various degrees (Vymazal et al., 2007). During the 20th century, constructed wetland technology has spread to every continent and is used both to treat sewage, gray water, stormwater overflows, and runoff and to recover nutrients and recycle the water within cities as an integrated part of circular urban drainage systems (Vymazal, 2022). Reed for treatment wetlands is usually locally produced, but new stands can be easily established by layering of green shoots on the ground (Véber, 1978). Non-native reed lineages, harvested alive from natural wetlands, could be used to establish new wastewater treatment enclosures. Both non-native and native lineages from North America have been shown to be highly efficient pollutant removers (Rodriguez & Brisson, 2015).

Economic Feasibility of Reed Use

Harvesting reed does not threaten the nature and the natural functions of wetlands. An economic large-scale harvest of wetland plants such as reed requires efficient machines that are adapted to saturated organic soils by having a low ground pressure. Machines equipped with balloon tires, modified snow groomers, and newly developed special-purpose tracked machinery are used for marsh biomass harvest (Joosten et al., 2012; Wichtmann et al., 2016). The cost-effectiveness of harvesting reeds is under-studied, and influences on profitability are not well-known. Wichmann (2017) suggested that the least profitable option for reed-use is harvest for biogas, with harvest of bales for biocombustion being more profitable, and use for construction (thatching) being highly profitable. Reed as a renewable energy source is potentially most useful for power generation in cofiring applications, as supplement in pellet form, due to the high bulk density (Croon, 2014).

The cost of constructed wetlands depends on the size of the wastewater treatment area and on the specific design. A direct comparison of costs for wetland remediation and conventional remediation would involve process design considerations for each input condition due to the case-specific nature of wastewater treatment and is therefore not readily available, let alone specifically for constructive wetlands using only reed as planted vegetation. A study comparing conventional (chemical) remediation costs with wetland remediation of wastewater found that treatment wetlands are more economically feasible (up to 25% cheaper) at small flow rates, on small scales, and for integrated secondary and tertiary treatment for phosphorus and nitrogen removal (Firth et al., 2020). To be a commercially attractive feedstock for non-wood pulp production, reed should ideally grow in monoculture on an area of 25,000–35,000 hectares, at a delivery cost at the mill of no more than $70 per ton. Thus, a—acceptable (for the pulp industry)—return on investment of 15% could be achieved (Croon, 2014).

Paludiculture as Example of Common Reed Cultivation on Rewetted Organic Soil

An increasing demand for arable land threatens pristine wetlands and organic soils in many parts of the world. Preservation of peat stock and sequestered organic carbon is important to ensure carbon storage, water and nutrient retention, and minimization of greenhouse gas emissions. Drainage of organic land results in the degradation of the soil and elevated CO_2 emissions due to aeration. Paludiculture (derived from the Latin term *palus*, meaning swamp or marsh) is the cultivation of rewetted peatland and organic soils, combining reduced greenhouse gas emissions and sustainable use options for the wet crops with the economic benefits of continued farming on the water-saturated soil (Joosten et al., 2012; Wichtmann et al., 2016). Paludiculture as alternative cultivation method on rewetted organic soil is gaining momentum. The UN Food and Agricultural Organisation (FAO) has acknowledged paludiculture as a sustainable activity to avoid rendering otherwise productive land useless because of ongoing subsidence and degradation. The Ramsar Convention on Wetlands mentions paludiculture as a sustainable option for the production of biofuels, and the Intergovernmental Panel on Climate Change (IPCC) has produced new guidelines that explicitly mention paludiculture as a land use option for re-wetted areas (Wichtmann & Couwenberg, 2013).

Species such as alder, *Sphagnum* sp., cattail, or reed canary grass are examples for potential, traditional and tested paludicrops, which are utilized as growing media, fuel, or for construction (Abel et al., 2013; Abel & Kallweit, 2022). *Phragmites australis* is an exemplary crop for paludiculture, since it is already widely used in large parts of Europe. Its longevity, harvestability, productivity, response to cutting, plant plasticity, and biomass quality have been evaluated to be especially valuable for paludiculture, in particular to be used for biogas conversion (Silvestri et al., 2017). Compared to the bioenergy crop *Miscanthus*, common reed yields less biomethane, but if biomass is to be produced sustainably, to preserve organic matter, and reduce greenhouse gas emissions, reed cultivated through paludiculture is a highly suitable plant (Dragoni et al., 2017; Ren et al., 2019). Its biomass production has been shown to be even more stimulated if fertilizer is added, although this is neither necessary nor desirable in eutrophicated systems, and in Europe even illegal on water-saturated soils (Vroom et al., 2022).

Cultivating invasive *Phragmites australis* as paludicrop in North America may not be an intuitive means of resource use. Nonetheless, the experience from European paludiculture and reed utilization can certainly be an inspiration for harvesting and thereby partially controlling invasive common reed as a renewable resource on the land that the species has inhabited.

Non-native Common Reed Utilization in North America—The Present and the Future

Utilization of invasive common reed in North America could be a sustainable alternative to attempts at merely eradicating the plant. Indeed, although common reed in Europe grows naturally prolifically in fresh and saline wetlands (Eller et al., 2017), floodplain modification has led to widespread reedbed monoculture in some Eastern European places, which displaces other native species and the reeds there are considered invasive (Goriup et al., 2019). In such affected floodplains of the rivers Prut, Danube, and Dniester in Ukraine, Moldova, and Romania, common reed is harvested for export as thatching bundles, some companies collect biomass for pellets and briquettes for boiler fuel, and a large part of the biomass is burned to revive the more open and dynamic habitats (Goriup et al., 2019). In the described area, some 100,000 tons of dry reed biomass per year could be produced as renewable biomass, equivalent to ~79,000 tons of CO_2 saved from burning fossil fuels (Goriup et al., 2019).

A promising outlook for invasive common reed biomass as bioenergy feedstock is the successful application of non-native *Phragmites australis* from Canada as biofuel pellets, which has previously been shown to be similar in quality to the bioenergy crop switchgrass (*Panicum virgatum*) (Vaicekonyte et al., 2013). It is also likely useful for biogas yield, as the intraspecific variation of the species seemingly plays no role in its biomethane production, and native European *Phragmites australis* produces fairly high CH_4 yields compared to other biofuel species (Eller et al., 2020a). Invasive common reed biomass (collected in Canada) is also suitable for bioethanol and even lignocellulosic butanol production after an adequate pretreatment; its production rate is comparable to the potential bioenergy crop switchgrass (Gao et al., 2014). The biomass yield of invasive common reed is about 18.6 tons per hectare in brackish tidal marshes in the U.S., a yield which is probably even higher in eutrophic, warm wetlands (Windham, 2001; Melts et al., 2019). Shoot height and aboveground tissue mass of non-native EU type reeds were among the highest when compared to other European and Asian common reed populations, and are therefore suitable for biomass harvest with high yields (Eller et al., 2020b).

Both the native and invasive lineages of *Phragmites australis* in North America have been shown to be appropriate in wastewater treatment wetlands, with comparable levels of pollutant removal efficiency (Rodríguez & Brisson, 2015). However, since common reed is considered an invasive plant species by natural resource and wildlife agencies, the use of this species has been limited in the United States (Wallace & Knight, 2006). Livestock grazing has the potential to reduce the cover of invasive common reed over the longer term and to lead to an improved ecological state of the grazed area, while providing an adequate feed with high crude protein content, although energy supplementation is recommended due to low in vitro dry matter digestibility (Brundage, 2010; Silliman et al., 2014; Volesky et al., 2016).

Historically, common reed played a prominent role for Native North American tribes, who used the plant for food, hunting and fishing devices, mats, baskets, cordage, musical instruments, shelter construction, clothing, ceremonial equipment, medicine, and more. Clearly, the current level of direct use is lower than historic and prehistoric use of *Phragmites australis*, especially in the western U.S. and northwestern Mexico, though its traditional applications were comparable to non-industrial use of the species in other parts of the world (Kiviat & Hamilton, 2001). Common reed still has many, not always visible functions, which is true also for invasive common reed in North America (Kiviat, 2013). Its high productivity can be exploited commercially by using the species' aboveground biomass as renewable energy resources or renewable raw material, to conserve nature areas, to minimize fossil fuel burning, and to partly control the spread of the invader. Thereby, the wetland function is preserved by allowing for re-growth in the following season and removal of excess nutrients from the soil is enhanced

(Alsbury, 2010). Existing assessments of necessary conditions for the commercial exploitation of common reed can help facilitate industrial use of its biomass also in North America (Croon, 2014; Tanneberger et al., 2021).

Literature Cited

Abel, S. & T. Kallweit. 2022. *Potential Paludiculture Plants of the Holarctic.* Proceedings of the Greifswald Mire Centre 04/2022. Greifswald Mire Centre, Greifswald.

Abel, S., J. Couwenberg, T. Dahms & H. Joosten. 2013. "The Database of Potential Paludiculture Plants (DPPP) and results for Western Pomerania." *Plant Diversity and Evolution* 130: 219–228.

Al-Haj, A. N. & R. W. Fulweiler. 2020. "A synthesis of methane emissions from shallow vegetated coastal ecosystems." *Global Change Biology* 26: 2988–3005.

Alsbury, S. 2010. *Stodmarsh Sustainable Reed Cutting.* RSPB, Wareham.

Andersen, L. H., P. Nummi, S. Bahrndorff, C. Pertoldi, K. Trøjelsgaard, T. L. Lauridsen, J. Rafn, C. M. S. Frederiksen, M. P. Kristjansen & D. Bruhn. 2021a. "Reed bed vegetation structure and plant species diversity depend on management type and the time period since last management." *Applied Vegetation Science* 24: e12531.

Andersen, L. H., P. Nummi, J. Rafn, C. M. S. Frederiksen, M. P. Kristjansen, T. L. Lauridsen, K. Trøjelsgaard, C. Pertoldi, D. Bruhn & S. Bahrndorff. 2021b. "Can reed harvest be used as a management strategy for improving invertebrate biomass and diversity?" *Journal of Environmental Management* 300: 113637.

Armstrong, J., F. Afreen-Zobayed & W. Armstrong. 1996a. "*Phragmites* die-back: Sulphide- and acetic acid-induced bud and root death, lignifications, and blockages within aeration and vascular systems." *New Phytologist* 134: 601–614.

Armstrong, J, W. Armstrong, P. M. Beckett, J. E. Halder, S. Lythe, R. Holt & A. Sinclair. 1996b. "Pathways of aeration and the mechanisms and beneficial effects of humidity- and Venturi-induced convections in *Phragmites australis* (Cav.) Trin. ex Steud." *Aquatic Botany* 54: 177–197.

Asaeda, T., L. Rajapakse, J. Manatunge & N. Sahara. 2006. "The effect of summer harvesting of *Phragmites australis* on growth characteristics and rhizome resource storage." *Hydrobiologia* 553: 327–335.

Back, C. L, J. R. Holomuzki, D. M. Klarer & R. S. Whyte. 2012. "Herbiciding invasive reed: Indirect effects on habitat conditions and snail-algal assemblages one year post-application." *Wetlands Ecology and Management* 20: 419–431.

Bakker, E. S., C. G. F. Veen, G. J. N. Ter Heerdt, N. Huig & J. M. Sarneel. 2018. "High grazing pressure of geese threatens conservation and restoration of reed belts." *Frontiers in Plant Science* 9: 1649.

Baran, M., Z. Varadyova, S. Kracmar & J. Hedbavny. 2002. "The common reed (*Phragmites australis*) as a source of roughage in ruminant nutrition." *Acta Veterinaria Brno* 71: 445–449.

Bartlett, K. B. & R. C. Harriss. 1993. "Review and assessment of methane emissions from wetlands." *Chemosphere* 26: 261–320.

Baute, K., L. L. Van Eerd, D. E. Robinson, P. H. Sikkema, M. Mushtaq & B. H. Gilroyed. 2018. "Comparing the biomass yield and biogas potential of *Phragmites australis* with *Miscanthus × giganteus* and *Panicum virgatum* grown in Canada." *Energies* 11: 2198.

Bavithra, G., J. Azevedo, F. Oliveira, J. Morais, E. Pinto, I. M. P. L. O. Ferreira, V. Vasconcelos, A. Campos & C. M. R. Almeida. 2020. "Assessment of constructed wetlands' potential for the removal of cyanobacteria and microcystins (MC-LR)." *Water* 12(1): 10.

Bernhardt, K. G. & M. Koch. 2003. "Restoration of a salt marsh system: Temporal change of plant species diversity and composition." *Basic and Applied Ecology* 4: 441–451.

Blossey, B., S. B. Endriss, R. Casagrande, P. Häfliger, H. Hinz, A. Dávalos, C. Brown-Lima, L. Tewksbury & R. S. Bourchier. 2020. "When misconceptions impede best practices: Evidence supports biological control of invasive *Phragmites*." *Biological Invasions* 22: 873–883.

Borin, M., G. Bonaiti, G. Santamaria & L. Giardini. 2001. "A constructed surface flow wetland for treating agricultural waste waters." *Water Science and Technology* 44: 523–530.

Bouza, H. & S. Asut. 2020. "Advancing reed-based architecture through circular digital fabrication." Pp. 117–126 *in* L. C. Werner & D. Koering (editors), *Anthropologic—Architecture and Fabrication in the Cognitive Age*, Vol. 1. Proceedings of the 38th eCAADe Conference on Education and Research in Computer Aided Architectural Design in Europe. eCAADe, Berlin.

Brix, H. 1987. "Treatment of wastewater in the rhizosphere of wetland plants—The root-zone method." *Water Science and Technology* 19: 107–118.

Brix, H. 1994. "Functions of macrophytes in constructed wetlands." *Water Science and Technology* 29: 71–78.

Brix, H. 1997. "Do macrophytes play a role in constructed treatment wetlands?" *Water Science and Technology* 35: 11–17.

Brix, H., B. K. Sorrell & H.-H. Schierup. 1996. "Gas fluxes achieved by in situ convective flow in *Phragmites australis*." *Aquatic Botany* 54: 151–163.

Brix, H., B. K. Sorrell & B. Lorenzen. 2001. "Are *Phragmites*-dominated wetlands a net source or net sink of greenhouse gases?" *Aquatic Botany* 69: 313–324.

Brix, H., S. Ye, E. A. Laws, D. Sun, G. Li, X. Ding, H. Yuan, G. Zhao, J. Wang & S. Pei. 2014. "Large-scale management of common reed, *Phragmites australis*, for paper production: A case study from the Liaohe Delta, China." *Ecological Engineering* 73: 760–769.

Brundage, A. 2010. *Grazing as a Management Tool for Controlling Phragmites australis and Restoring Native Plant Biodiversity in Wetlands*. M.S. Thesis, University of Maryland, College Park.

Bumby, C. & E. C. Farrer. 2022. "*Nipponaclerda biwakoensis* infestation of *Phragmites australis* in the Mississippi River Delta, USA: Do fungal microbiomes play a role?" *Wetlands* 42: 15.

Burdick, D. M., R. Buchsbaum & E. Holt. 2001. "Variation in soil salinity associated with expansion of *Phragmites australis* in salt marshes." *Environmental and Experimental Botany* 46: 247–261.

Calheiros, C. S. C., A. O. S. S. Rangel & P. M. L. Castro. 2009. "Treatment of industrial wastewater with two-stage constructed wetlands planted with *Typha latifolia* and *Phragmites australis*." *Bioresource Technology* 100: 3205–3213.

Cavalaglio, G., M. Gelosia, D. Ingles, E. Pompili, S. D'Antonio & F. Cotana. 2016. "Response surface methodology for the optimization of cellulosic ethanol production from *Phragmites australis* through pre-saccharification and simultaneous saccharification and fermentation." *Industrial Crops and Products* 83: 431–437.

Chambers, R. M., L. A. Meyerson & K. Saltonstall. 1999. "Expansion of *Phragmites australis* into tidal wetlands of North America." *Aquatic Botany* 64: 261–273.

Chambers, R. M., D. T. Osgood, D. J. Bart & F. Montalto. 2003. "*Phragmites australis* invasion and expansion in tidal wetlands: Interactions among salinity, sulfide, and hydrology." *Estuaries* 26: 398–406.

Chivu, A. I. 1968. "Practical experiment in the cropping of reeds for the manufacture of pulp and paper—Economic results." Pp. 877–899 *in* FAO (editor), *Pulp and Paper Development in Africa and the Near East*, Vol. 2. Food and Agriculture Organization of the United Nations (FAO), Rome.

Čížková, H., H. Brix, J. Kopecký & J. Lukavská. 1999. "Organic acids in the sediments of wetlands dominated by *Phragmites australis*: Evidence of phytotoxic concentrations." *Aquatic Botany* 64: 303–315.

Clevering, O. A. 1999. "The effects of litter on growth and plasticity of *Phragmites australis* clones originating from infertile, fertile or eutrophicated habitats." *Aquatic Botany* 64: 35–50.

Clevering, O. A. & J. Lissner. 1999. "Taxonomy, chromosome numbers, clonal diversity and population dynamics of *Phragmites australis*." *Aquatic Botany* 64: 185–208.

Connor, H. E., M. I. Dawson, R. D. Keating & L. S. Gill. 1998. "Chromosome numbers of *Phragmites australis* (Arundineae: Gramineae) in New Zealand." *New Zealand Journal of Botany* 36: 465–469.

Cronin, J. T., G. P. Bhattarai, W. J. Allen & L. A. Meyerson. 2015. "Biogeography of a plant invasion: Plant–herbivore interactions." *Ecology* 96: 1115–1127.

Cronin, J. T., E. Kiviat, L. A. Meyerson, G. P. Bhattarai & W. J. Allen. 2016. "Biological control of invasive *Phragmites australis* will be detrimental to native *P. australis*." *Biological Invasions* 18: 2749–2752.

Cronin, J. T., J. Johnston & R. Diaz. 2020. "Multiple potential stressors and dieback of *Phragmites australis* in the Mississippi River Delta, USA: Implications for restoration." *Wetlands* 40: 2247–2261.

Croon, F. W. 2014. "Saving reed lands by giving economic value to reed." *Mires and Peat* 13(10).

Czubaszek, R., A. Wysocka-Czubaszek, S. Roj-Rojewski & P. Banaszuk. 2019. "Greenhouse gas fluxes from soils fertilised with anaerobically digested biomass from wetlands." *Mires and Peat* 25(Apr. 1): 1–11.

Davidson, A. M., M. Jennions & A. B. Nicotra. 2011. "Do invasive species show higher phenotypic plasticity than native species and, if so, is it adaptive? A meta-analysis." *Ecology Letters* 14: 419–431.

Den Hartog, C., J. Květ & H. Sukopp. 1989. "Reed. A common species in decline." *Aquatic Botany* 35: 1–4.

Derr, J. F. 2008. "Common reed (*Phragmites australis*) response to mowing and herbicide application." *Invasive Plant Science and Management* 1: 12–16.

Dioha, I. J., C. H. Ikeme, T. Nafi'u & N. I. Soba. 2013. "Effect of carbon to nitrogen ratio on biogas production." *International Research Journal of Natural Sciences* 1(3): 1–10.

Dragoni, F., V. Giannini, G. Ragaglini, E. Bonari & N. Silvestri. 2017. "Effect of harvest time and frequency on biomass quality and biomethane potential of common reed (*Phragmites australis*) under paludiculture conditions." *BioEnergy Research* 10: 1066–1078.

Dukes, J. S. & H. A. Mooney. 1999. "Does global change increase the success of biological invaders?" *Trends in Ecology and Evolution* 14: 135–139.

Durant, D., A. Farruggia & A. Tricheur. 2020. "Utilization of common reed (*Phragmites australis*) as bedding for housed suckler cows: Practical and economic aspects for farmers." *Resources* 9: 140.

Eller, F., C. Lambertini, L. X. Nguyen & H. Brix. 2014. "Increased invasive potential of non-native *Phragmites australis*: Elevated CO_2 and temperature alleviate salinity effects on photosynthesis and growth." *Global Change Biology* 20: 531–543.

Eller, F., H. Skálová, J. S. Caplan, G. P. Bhattarai, M. K. Burger, J. T. Cronin, W.-Y. Guo, et al. 2017. "Cosmopolitan species as models for ecophysiological responses to global change: The common reed *Phragmites australis*." *Frontiers in Plant Science* 8: 1833.

Eller, F., P. M. Ehde, C. Oehmke, L. Ren, H. Brix, B. K. Sorrell & S. E. B. Weisner. 2020a. "Biomethane yield from different European *Phragmites australis* genotypes, compared with other herbaceous wetland species grown at different fertilization regimes." *Resources* 9: 57.

Eller, F., X. Guo, S. Ye, T. J. Mozdzer & H. Brix. 2020b. "Suitability of wild *Phragmites australis* as bio-resource: Tissue quality and morphology of populations from three continents." *Resources* 9: 143.

Emery, H. E. & R. W. Fulweiler. 2014. "*Spartina alterniflora* and invasive *Phragmites australis* stands have similar greenhouse gas emissions in a New England marsh." *Aquatic Botany* 116: 83–92.

Firth, A. E. J., N. M. Dowell, P. S. Fennell & J. P. Hallett. 2020. "Assessing the economic viability of wetland remediation of wastewater, and the potential for parallel biomass valorisation." *Environmental Science: Water Research & Technology* 6: 2103–2121.

Fujibayashi, M., F. Takakai, S. Masuda, K. Okano & N. Miyata. 2020. "Effects of restoration of emergent macrophytes on the benthic environment of the littoral zone of a eutrophic lake." *Ecological Engineering* 155: 105960.

Gao, K., S. Boiano, A. Marzocchella & L. Rehmann. 2014. "Cellulosic butanol production from alkali-pretreated switchgrass (*Panicum virgatum*) and phragmites (*Phragmites australis*)." *Bioresource Technology* 174: 176–181.

García-Valero, A., S. Martínez-Martínez, Á. Faz, M. A. Terrero, M. Á. Muñoz, M. D. Gómez-López & J. A. Acosta. 2020. "Treatment of wastewater from the tannery industry in a constructed wetland planted with *Phragmites australis*." *Agronomy* 10: 176.

Giannini, V., C. Oehmke, N. Silvestri, W. Wichtmann, F. Dragoni & E. Bonari. 2016. "Combustibility of biomass from perennial crops cultivated on a rewetted Mediterranean peatland." *Ecological Engineering* 97: 157–169.

Gigante, D., C. Angiolini, F. Landucci, F. Maneli, B. Nisi, O. Vaselli, R. Venanzoni & L. Lastrucci. 2014. "New occurrence of reed bed decline in southern Europe: Do permanent flooding and chemical parameters play a role?" *Comptes Rendus Biologies* 337: 487–498.

Gorenflot, R., J.-M. Hubac, M. Jay & P. Lalande. 1993. "Geographic distribution, polyploidy and pattern of flavonoids in *Phragmites australis* (Cav.) Trin. ex Steud." Pp. 474–478 *in* J. Felsenstein (editor), *Numerical Taxonomy*. NATO ASI Series G, Vol. 1. Springer-Verlag Berlin, Heidelberg.

Goriup, P., A. Haberl, O. Rubel, V. Ajder, I. Kulchytskyy, A. Smaliychuk & N. Goriup. 2019. "Potential for renewable use of biomass from reedbeds on the lower Prut, Danube and Dniester floodplains of Ukraine and Moldova." *Mires and Peat* 25 (Sept. 16).

Günther, A., A. Barthelmes, V. Huth, H. Joosten, G. Jurasinski, F. Koebsch & J. Couwenberg. 2020. "Prompt rewetting of drained peatlands reduces climate warming despite methane emissions." *Nature Communications* 11: 1644.

Guo, W. Y., C. Lambertini, X. Z. Li, L. A. Meyerson & H. Brix. 2013. "Invasion of Old World *Phragmites australis* in the New World: Precipitation and temperature patterns combined with human influences redesign the invasive niche." *Global Change Biology* 19: 3406–3422.

Güsewell, S. 2003. "Management of *Phragmites australis* in Swiss fen meadows by mowing in early summer." *Wetlands Ecology and Management* 11: 433–445.

Güsewell, S., A. Buttler & F. Klötzli. 1998. "Short-term and long-term effects of mowing on the vegetation of two calcareous fens." *Journal of Vegetation Science* 9: 861–872.

Hansson, P.-A. & H. Fredriksson. 2004. "Use of summer harvested common reed (*Phragmites australis*) as nutrient source for organic crop production in Sweden." *Agriculture, Ecosystems & Environment* 102: 365–375.

Hauber, D. P., K. Saltonstall, D. A. White & C. S. Hood. 2011. "Genetic variation in the common reed, *Phragmites australis*, in the Mississippi River Delta marshes: Evidence for multiple introductions." *Estuaries and Coasts* 34: 851–862.

Hazelton, E. L. G., T. J. Mozdzer, D. M. Burdick, K. M. Kettenring & D. F. Whigham. 2014. "*Phragmites australis* management in the United States: 40 years of methods and outcomes." *AoB PLANTS* 6: plu001.

Hellmann, J. J., J. E. Byers, B. G. Bierwagen & J. S. Dukes. 2008. "Five potential consequences of climate change for invasive species." *Conservation Biology* 22: 534–543.

Jasinskas, A., D. Streikus, E. Šarauskis, M. Palšauskas & K. Venslauskas. 2020. "Energy evaluation and greenhouse gas emissions of reed plant pelletizing and utilization as solid biofuel." *Energies* 13: 1516.

Jenkins, B. M., L. L. Baxte, T. R. Miles Jr. & T. R. Miles. 1998. "Combustion properties of biomass." *Fuel Processing Technology* 54: 17–46.

Jensen, J. K. 2020. *Naturens Eget Tag*. Stråtagets Kontor, Aarhus.

Jeschke, L. 1987. "Vegetationsdynamik des Salzgraslandes im Bereich der Ostseeküste der DDR unter Einfluß des Menschen" ["Anthropic vegetation dynamics of salt grasslands at the Baltic Sea shore of the GDR"]. *Hercynia* 24: 321–328.

Joosten, H., M. L. Tapio-Bistrom & S. Tol (editors). 2012. *Peatlands—Guidance for Climate Change Mitigation Through Conservation, Rehabilitation and Sustainable Use.* Second edition. Mitigation of Climate Change in Agriculture 5. Food and Agriculture Organization of the United Nations (FAO) and Wetlands International, Rome.

Jurasinski, G., S. Ahmad, A. Anadon-Rosell, J. Berendt, F. Beyer, R. Bill, G. Blume-Werry, et al. 2020. "From understanding to sustainable use of peatlands: The WETSCAPES approach." *Soil Systems* 4: 14.

Karlsson, N. P. E., F. Halila, M. Mattsson & M. Hoveskog. 2017. "Success factors for agricultural biogas production in Sweden: A case study of business model innovation." *Journal of Cleaner Production* 142: 2925–2934.

Kask, U., L. Kask & S. Link. 2013. "Combustion characteristics of reed and its suitability as a boiler fuel." *Mires and Peat* 13(5): 1–10.

Kickuth, R. 1980. "Abwasserbehandlung im Wurzelraumverfahren." *WLB Wasser, Luft und Betrieb* 11: 21–24.

Kiviat, E. 2013. "Ecosystem services of *Phragmites* in North America with emphasis on habitat functions." *AoB PLANTS* 5: plt008.

Kiviat, E. 2019. "Organisms using *Phragmites australis* are diverse and similar on three continents." *Journal of Natural History* 53: 1975–2010.

Kiviat, E. & E. Hamilton. 2001. "*Phragmites* use by Native North Americans." *Aquatic Botany* 69: 341–357.

Kiviat, E., L. A. Meyerson, T. J. Mozdzer, W. J. Allen, A. H. Baldwin, G. P. Bhattarai, H. Brix, et al. 2019. "Evidence does not support the targeting of cryptic invaders at the subspecies level using classical biological control: The example of *Phragmites*." *Biological Invasions* 21: 2529–2541.

Knight, I. A., B. E. Wilson, M. Gill, L. Aviles, J. T. Cronin, J. A. Nyman, S. A. Schneider & R. Diaz. 2018. "Invasion of *Nipponaclerda biwakoensis* (Hemiptera: Aclerdidae) and *Phragmites australis* die-back in southern Louisiana, USA." *Biological Invasions* 20: 2739–2744.

Knight, I. A., J. T. Cronin, M. Gill, J. A. Nyman, B. E. Wilson & R. Diaz. 2020. "Investigating plant phenotype, salinity, and infestation by the Roseau Cane Scale as factors in the die-back of *Phragmites australis* in the Mississippi River Delta, USA." *Wetlands* 40: 1327–1337.

Knight, R. L., R. H. Kadlec & H. M. Ohlendorf. 1999. "The use of treatment wetlands for petroleum industry effluents." *Environmental Science & Technology* 33: 973–980.

Köbbing, J. F., N. Thevs & S. Zerbe. 2013. "The utilisation of reed (*Phragmites australis*): A review." *Mires and Peat* 13(1): 1–14.

Kovács, M. G., G. Turcsányi, Z. D. Tuba, S. E. Wolcsanszky, T. Vásárhelyi, A. D. Draskovits, S. Tóth, A. Koltay & L. Kaszab. 1989. "The decay of reed in Hungarian lakes." *Symposia Biologica Hungarica* 38: 461–471.

Kvet, J., J. Dusek, S. Husak & J. Vymazal. 1999. "Vascular plants suitable for wastewater treatment in temperate zones." Pp. 101–110 *in* J. Vymazal (editor), *Nutrient Cycling and Retention in Natural and Constructed Wetlands.* Backhuys Publishers, Leiden.

Lambertini, C., I. A. Mendelssohn, M. H. G. Gustafsson, B. Olesen, T. Riis, B. K. Sorrell & H. Brix. 2012. "Tracing the origin of Gulf Coast *Phragmites* (Poaceae): A story of long-distance dispersal and hybridization." *American Journal of Botany* 99: 538–551.

Lambertini, C., W.-Y. Guo, S. Ye, F. Eller, X. Guo, X.-Z. Li, B. K. Sorrell, M. Speranza & H. Brix. 2020. "Phylogenetic diversity shapes salt tolerance in *Phragmites australis* estuarine populations in East China." *Scientific Reports* 10: 17645.

Lewandowski, I. & D. Kauter. 2003. "The influence of nitrogen fertilizer on the yield and combustion quality of whole grain crops for solid fuel use." *Industrial Crops and Products* 17: 103–117.

Link, S., U. Kask, J. Krail & H. Plank. 2015. "Combustion tests of reed pellets." Pp. 697–701 *in* I. Obernberger, D. Baxter, A. Grassi & P. Helm (editors), *Setting the Course for a Biobased Economy*. Papers of the 23rd European Biomass Conference.

Liu, L.-L., M.-Q. Yin, X. Guo, J.-W. Wang, Y.-F. Cai, C. Wang, X.-N. Yu, et al. 2022. "Cryptic lineages and potential introgression in a mixed-ploidy species (*Phragmites australis*) across temperate China." *Journal of Systematics and Evolution* 60: 398–410.

Lizasoain, J., M. Rincón, F. Theuretzbacher, R. Enguídanos, P. J. Nielsen, A. Potthast, T. Zweckmair, A. Gronauer & A. Bauer. 2016. "Biogas production from reed biomass: Effect of pretreatment using different steam explosion conditions." *Biomass and Bioenergy* 95: 84–91.

Lu, M., J. S. Caplan, J. D. Bakker, J. A. Langley, T. J. Mozdzer, B. G. Drake & J. P. Megonigal. 2016. "Allometry data and equations for coastal marsh plants." *Ecology* 97: 3554.

Martin, R. M. & S. Moseman-Valtierra. 2015. "Greenhouse gas fluxes vary between *Phragmites Australis* and native vegetation zones in coastal wetlands along a salinity gradient." *Wetlands* 35: 1021–1031.

Martin, R. M. & S. Moseman-Valtierra. 2017. "Effects of transient *Phragmites australis* removal on brackish marsh greenhouse gas fluxes." *Atmospheric Environment* 158: 51–59.

Mayer, F., P. A. Gerin, A. Noo, S. Lemaigre, D. Stilmant, T. Schmit, N. Leclech, et al. 2014. "Assessment of energy crops alternative to maize for biogas production in the Greater Region." *Bioresource Technology* 166: 358–367.

Melts, I., M. Ivask, M. Geetha, K. Takeuchi & K. Heinsoo. 2019. "Combining bioenergy and nature conservation: An example in wetlands." *Renewable and Sustainable Energy Reviews* 111: 293–302.

Meyerson, L. A., K. Saltonstall, L. Windham, E. Kiviat & S. Findlay. 2000. "A comparison of *Phragmites australis* in freshwater and brackish marsh environments in North America." *Wetlands Ecology and Management* 8: 89–103.

Meyerson, L. A., A. M. Lambert & K. Saltonstall. 2010. "A tale of three lineages: Expansion of common reed (*Phragmites australis*) in the U.S. Southwest and Gulf Coast." *Invasive Plant Science and Management* 3: 515–520.

Milani, M. & A. Toscano. 2013. "Evapotranspiration from pilot-scale constructed wetlands planted with *Phragmites australis* in a Mediterranean environment." *Journal of Environmental Science and Health*, Part A 48: 568–580.

Miljan, M., M.-J. Miljan, J. Miljan, K. Akermann & K. Karja. 2013. "Thermal transmittance of reed-insulated walls in a purpose-built test house." *Mires and Peat* 13(7): 1–12.

Monjeau, C. 1901. Purifying Water. U.S. Patent 681,884, 18 December 1901.

Monllor, P., C. A. Sandoval-Castro, A. J. Ayala-Burgos, A. Roca, G. Romero & J. R. Díaz. 2020. "Preference study of four alternative silage fodders from the Mediterranean region in Murciano-Granadina goats." *Small Ruminant Research* 192: 106204.

Moore, G. E., D. M. Burdick, C. R. Peter & D. R. Keirstead. 2012. "Belowground biomass of *Phragmites australis* in coastal marshes." *Northeastern Naturalist* 19: 611–626.

Mozdzer, T. J. & J. C. Zieman. 2010. "Ecophysiological differences between genetic lineages facilitate the invasion of non-native *Phragmites australis* in North American Atlantic coast wetlands." *Journal of Ecology* 98: 451–458.

Mozdzer, T. J. & J. P. Megonigal. 2012. "Jack-and-master trait responses to elevated CO_2 and N: A comparison of native and introduced *Phragmites australis*." *PLoS ONE* 7: e42794.

Mozdzer, T. J. & J. P. Megonigal. 2013. "Increased methane emissions by an introduced *Phragmites australis* lineage under global change." *Wetlands* 33: 609–615.

Mozdzer, T. J., J. C. Zieman & K. J. McGlathery. 2010. "Nitrogen uptake by native and invasive temperate coastal macrophytes: Importance of dissolved organic nitrogen." *Estuaries and Coasts* 33: 784–797.

Mueller, P., R. N. Hager, J. E. Meschter, T. J. Mozdzer, J. A. Langley, K. Jensen & J. P. Megonigal. 2016. "Complex invader-ecosystem interactions and seasonality mediate the impact of non-native *Phragmites* on CH_4 emissions." *Biological Invasions* 18: 2635–2647.

Ostendorp, W. 1989. "'Die-back' of reeds in Europe—A critical review of literature." *Aquatic Botany* 35: 5–26.

Ranieri, E., P. Verlicchi & T. M. Young. 2011. "Paracetamol removal in subsurface flow constructed wetlands." *Journal of Hydrology* 404: 130–135.

Ren, L., F. Eller, C. Lambertini, W. Y. Guo, H. Brix & B. K. Sorrell. 2019. "Assessing nutrient responses and biomass quality for selection of appropriate paludiculture crops." *Science of the Total Environment* 664: 1150–1161.

Richards, C. L., O. Bossdorf, N. Z. Muth, J. Gurevitch & M. Pigliucci. 2006. "Jack of all trades, master of some? On the role of phenotypic plasticity in plant invasions." *Ecology Letters* 9: 981–993.

Rodríguez, M. & J. Brisson. 2015. "Pollutant removal efficiency of native versus exotic common reed (*Phragmites australis*) in North American treatment wetlands." *Ecological Engineering* 74: 364–370.

Rohal, C. B., C. Cranney, E. L. G. Hazelton & K. M. Kettenring. 2019. "Invasive *Phragmites australis* management outcomes and native plant recovery are context dependent." *Ecology and Evolution* 9: 13835–13849.

Rooth, J. E. & J. C. Stevenson. 2000. "Sediment deposition patterns in *Phragmites australis* communities: Implications for coastal areas threatened by rising sea-level." *Wetlands Ecology and Management* 8: 173–183.

Saltonstall, K. 2002. "Cryptic invasion by a non-native genotype of the common reed, *Phragmites australis*, into North America." *Proceedings of the National Academy of Sciences of the United States of America* 99: 2445–2449.

Schneider, S. A., H. J. Broadley, J. C. Andersen, J. S. Elkinton, S.-Y. Hwang, C. Liu, S. Noriyuki, et al. 2022. "An invasive population of Roseau Cane Scale in the Mississippi River Delta, USA originated from northeastern China." *Biological Invasions* 24: 2735–2755.

Sciance, M. B., C. J. Patrick, D. E. Weller, M. N. Williams, M. K. McCormick & E. L. G. Hazelton. 2016. "Local and regional disturbances associated with the invasion of Chesapeake Bay marshes by the common reed *Phragmites australis*." *Biological Invasions* 18: 2661–2677.

Seidel, K. 1953. "Pflanzungen zwischen gewassern und land." *Mitteilungen Max-Planck Gesselschaft* 8: 17–20.

Seidel, K. 1961. "Zur Problematik der Keim-und Pflanzgewässer: Mit 4 Abbildungen, 4 Tabellen und 4 Skizzen im Text." *Internationale Vereinigung für theoretische und angewandte Limnologie: Verhandlungen* 14: 1035–1043.

Seidel, K. 1964. "Abbau von bacterium coli durch höhere wasserpflanzen." *Naturwissenschaften* 51: 395.

Seidel, K. 1965. "Neue Wege zur Grundwasseranreicherung in Krefeld, Vol. II. Hydrobotanische Reinigungsmethode." *GWF Wasser/Abwasser* 30: 831–833.

Sheng, Y. P., A. A. Rivera-Nieves, R. Zou, V. A. Paramygin, C. Angelini & S. J. Sharp. 2021. "Invasive *Phragmites* provides superior wave and surge damage protection relative to native plants during storms." *Environmental Research Letters* 16: 54008.

Silliman, B. R., T. Mozdzer, C. Angelini, J. E. Brundage, P. Esselink, J. P. Bakker, K. B. Gedan, J. van de Koppel & A. H. Baldwin. 2014. "Livestock as potential biological control agent for an invasive wetland plant." *PeerJ* 2: e567.

Silvestri, N., V. Giannini, F. Dragoni & E. Bonari. 2017. "A multi-adaptive framework for the crop choice in paludicultural cropping systems." *Italian Journal of Agronomy* 12: 69–76.

Sweers, W., S. Horn, G. Grenzdoerffer & J. Mueller. 2013. "Regulation of reed (*Phragmites australis*) by water buffalo grazing: Use in coastal conservation." *Mires and Peat* 13(3): 1–10.

Szijártó, N., Z. Kádár, E. Varga, A. B. Thomsen, M. Costa-Ferreira & K. Réczey. 2009. "Pretreatment of reed by wet oxidation and subsequent utilization of the pretreated fibers for ethanol production." *Applied Biochemistry and Biotechnology* 155: 83–93.

Tanneberger, F., S. Abel, J. Couwenberg, T. Dahms, G. Gaudig, A. Günther, J. Kreyling, J. Peters, J. Pongratz & H. Joosten. 2021. "Towards net zero CO_2 in 2050: An emission reduction pathway for organic soils in Germany." *Mires and Peat* 27(5): 1–17.

Tóth, V. R. 2016. "Reed stands during different water level periods: Physico-chemical properties of the sediment and growth of *Phragmites australis* of Lake Balaton." *Hydrobiologia* 778: 193–207.

Tozluoglu, A. 2018. "Bioethanol production from common reed (*Phragmites australis*): Enzymatic hydrolysis and fermentation." *Romanian Biotechnological Letters* 23: 13473–13478.

Vaicekonyte, R., E. Kiviat, F. Nsenga & A. Ostfeld. 2013. "An exploration of common reed (*Phragmites australis*) bioenergy potential in North America." *Mires and Peat* 13(12): 1–9.

van der Putten, W. H. 1997. "Die-back of *Phragmites australis* in European wetlands: An overview of the European Research Programme on Reed Die-back and Progression (1993–1994)." *Aquatic Botany* 59: 263–275.

Véber, K. 1978. "Propagation, cultivation and exploitation of common reed in Czechoslovakia." Pp. 416–423 in D. Dykyjová & J. Květ (editors), *Pond Littoral Ecosystems*. Springer-Verlag Berlin, Heidelberg.

Volesky, J. D., S. L. Young & K. H. Jenkins. 2016. "Cattle grazing effects on *Phragmites australis* in Nebraska." *Invasive Plant Science and Management* 9: 121–127.

Vroom, R. J. E., J. J. M. Geurts, R. Nouta, A. C. W. Borst, L. P. M. Lamers & C. Fritz. 2022. "Paludiculture crops and nitrogen kick-start ecosystem service provisioning in rewetted peat soils." *Plant and Soil* 474: 337–354.

Vymazal, J. 2011. "Plants used in constructed wetlands with horizontal subsurface flow: A review." *Hydrobiologia* 674: 133–156.

Vymazal, J. 2013. "Emergent plants used in free water surface constructed wetlands: A review." *Ecological Engineering* 61: 582–592.

Vymazal, J. 2022. "The historical development of constructed wetlands for wastewater treatment." *Land* 11: 174.

Vymazal, J. & T. Březinová. 2016. "Accumulation of heavy metals in aboveground biomass of *Phragmites australis* in horizontal flow constructed wetlands for wastewater treatment: A review." *Chemical Engineering Journal* 290: 232–242.

Vymazal, J., J. Švehla, L. Kröpfelová & V. Chrastný. 2007. "Trace metals in *Phragmites australis* and *Phalaris arundinacea* growing in constructed and natural wetlands." *Science of the Total Environment* 380: 154–162.

Wallace, S. D. & R. L. Knight. 2006. *Small-Scale Constructed Wetland Treatment Systems: Feasibility, Design Criteria, and O&M Requirements*. IWA Publishing, London.

Wichmann, S. 2017. "Commercial viability of paludiculture: A comparison of harvesting reeds for biogas production, direct combustion, and thatching." *Ecological Engineering* 103: 497–505.

Wichmann, S. & J. F. Köbbing. 2015. "Common reed for thatching—A first review of the European market." *Industrial Crops and Products* 77: 1063–1073.

Wichtmann, W. & J. Couwenberg. 2013. "Reed as a renewable resource and other aspects of paludiculture." *Mires and Peat* 13(00): 1–2.

Wichtmann, W., C. Oehmke, S. Baerisch, F. Deschan, U. Malashevich & F. Tanneberger. 2013. "Combustibility of biomass from wet fens in Belarus and its potential as a substitute for peat in fuel briquettes." *Mires and Peat* 13(6): 1–10.

Wichtmann, W, C. Schroder & H. Joosten (editors). 2016. *Paludiculture—Productive Use of Wet Peatlands*. Schweizerbart Science Publishers, Stuttgart.

Windham, L. 2001. "Comparison of biomass production and decomposition between *Phragmites australis* (common reed) and *Spartina patens* (salt hay grass) in brackish tidal marshes of New Jersey, USA." *Wetlands* 21: 179–188.

Windham, L. & R. G. Lathrop. 1999. "Effects of *Phragmites australis* (common reed) invasion on aboveground biomass and soil properties in brackish tidal marsh of the Mullica river, New Jersey." *Estuaries* 22: 927.

Woehler-Geske, A., C. R. Moschner, A. Gellerich, H. Militz, J.-M. Greef & E. Hartung. 2016. "Provenances and properties of thatching reed (*Phragmites australis*)." *Landbauforschung* 66: 1–10.

Yacano, M. R., S. P. Thompson & M. F. Piehler. 2022. "Non-native marsh grass (*Phragmites australis*) enhances both storm and ambient nitrogen removal capacity in marine systems." *Estuaries and Coasts* 45: 2012–2025.

Zedler, J. B. & S. Kercher. 2004. "Causes and consequences of invasive plants in wetlands: Opportunities, opportunists, and outcomes." *Critical Reviews in Plant Sciences* 23: 431–452.

CHAPTER 10

Remediation as Harvest: Invasive Plant Species as Building Materials

Katie MacDonald & Kyle Schumann

In the American South, tales of invasive plant species loom large in the cultural imagination. In particular, Tennessee, a state which borders eight others, lies vulnerable to species migration from a variety of neighbors with assorted legislative approaches to addressing and controlling invasives. Leading offenders such as kudzu elicit particular condemnation from many parties, disgruntled homeowners and ecologists alike (Blaustein, 2001). Beyond some obvious offenders, an extensive list of some 40 established and 24 emerging species have been classified as posing immediate danger to local ecologies.[1]

Plant species regarded as invasive often arrive in new environments through human activities, both intentionally and unintentionally. Kudzu, for instance, was intentionally planted across the United States through

---

[1]. This unfortunate reality forms the basis for the work of the Tennessee Citizens for Wilderness Planning, which works to control exotic invasive species on public lands and trails, and the Tennessee Invasive Plant Council, which seeks to raise public awareness about the spread of invasive exotic plants and facilitate management and control of invasive exotic plants. These organizations facilitated efforts to secure safe, legal access to invasive plant specimens for the projects described in this chapter. https://www.tnipc.org/invasive-plants and https://tcwp.org/

government initiatives to enact erosion control methods along interstate roadways and across farmland (Finch, 2015). Once species find themselves in a new environment, further spread is facilitated by daily human activities. Roadways and waterways prove to be corridors for not just people and goods, but invasive plants that are transported along with items such as lumber, building products, landscape materials, and other goods (Harron et al., 2020).

The spread of invasive species usually portends ecological crises, as delicate ecosystems are thrown out of balance. Invasive plants are often characterized by fast growth rates that allow them to rapidly establish themselves and outcompete native species. The resilience of invasive species means that remediation efforts can be costly or nearly impossible. For example, kudzu (Fig. 10-1) has large tuber-like subterranean growths that can store energy, allowing the plant to go dormant underground for years before reemerging. Remediation methods involve an array of strategies, from mechanical removal and destruction (pulling, cutting, girdling) to chemical (herbicide) and even biological methods (insects and hungry goats). The cost burden for communities often outweighs economic benefits, creating substantial barriers to restoring native ecosystems. Instead, community organizations, non-profits, and volunteer activities represent the bulk of remediation efforts, and the limited resources of these collective groups represent barriers to extensive habitat restoration.

FIGURE 10-1. A common sight in the region: kudzu, "the vine that ate the South," overtaking buildings. Memphis, Tennessee, 2020. (Photo by Kevin Saslawsky.)

Material Disruption

Powerful adversaries, invasive species are instead considered as potential assets in this chapter. These plants are tenacious survivors whose physical material has been optimized through evolutionary processes to perform various functions. Here, we consider the characteristics of invasive plants in search of possible applications, with the goal that utility as carbon-sequestering construction materials may allow invasive plant material to gain value and, in turn, incentivize harvesting of this abundant material. In other words, the act of harvesting invasive plants can both reduce the abundance of these plants in the environment, making room for the restoration of native species and ecosystems, and provide value in the form of useful raw material for construction.

Since industrialization, construction has favored concrete and steel for their strength and longevity, materials which must now be reconsidered due to their embodied carbon and lifecycle costs. Wood offers advantages in terms of renewability and carbon sequestration, but presents limitations in growth cycle and resource availability. Material processing and transportation from forest to sawmill to jobsite result in significant embodied carbon, and much of the grown material is lost as logs are transformed into lumber. Furthermore, engineered wood products can leach chemicals and affect indoor air quality through off-gassing from adhesives and coatings. The farming, harvesting, and sourcing of timber poses environmental challenges: single-species cultivation, or monoculture, overcomes and destroys native habitats, acting as a kind of invasive at a macro scale, while multiple-species cultivation, or polyculture, provides an alternative practice that can benefit forest ecologies.

Unlike forestry practices, the sourcing of invasive plant species, as suggested here, does not require planting and farming. Instead, infested ecosystems are mined for source material. Invasive plant species sequester carbon at a range of rates, providing a superior carbon sink alternative to carbon-emitting construction materials such as concrete and steel.

In this chapter, two projects are discussed that demonstrate applications for three invasive plant species as construction materials. By finding uses for invasive species in design and construction, these projects incentivize the removal of such species, and in turn, the remediation of the environment at a regional scale. Moving beyond net-zero and carbon-negative building methods, this research pursues a symbiotic relationship between design and ecology—an ecology-positive architecture.

Homegrown

Homegrown (Figs. 10-2, 10-3) is a temporary installation that seeks to expand the inventory of carbon-sequestering grown materials used in construction and the processes allowing these materials to become building products. If logs are typically reduced into lumber through a *subtractive*, energy-intensive process, which produces waste in the form of offcuts, *Homegrown* takes an *additive* approach, aggregating small-scale plant species into a series of composite panels.

FIGURE 10-2. *Homegrown.* Knoxville Museum of Art, Knoxville, Tennessee, 2020. (Photo by authors.)

The installation brings together several invasive plant species and landscaping waste in a multi-species, composite material system. This method allows for the advantageous qualities of each species to create a cohesive product greater than the sum of its individual parts: the invasive vine kudzu is employed as a hearty, ropelike component that can weave and tangle through the other materials, helping adhesion in the tradition of jammed structures (Cohen et al., 2020); invasive golden bamboo provides a flexible linear component that can be bundled, woven, and layered; a multi-species collection of arched branches provide riblike elements that can be woven into the bamboo poles; and finally, pine needles provide a small aggregate for filling voids and providing an even exterior surface.

Homegrown draws techniques from existing wood composite materials and augments the possibilities by expanding the palette of material inputs and retooling the process to leverage the qualities of each species. Existing products composed of fine plant materials include oriented strand board (OSB), which is composed of compressed wood shards, and wood wool, which is composed of compressed wood noodles. In the fabrication of such products, trees are de-barked and run through a strander, producing chips which are then dried at high temperatures. The strands are mixed with resins and wax and laid up into panels, pressed at 600 PSI and heated at 425 degrees Fahrenheit to cure the adhesive resin. The resulting panels are rated for various applications in light timber construction, but drawbacks include the environmental impact and off-gassing of the resins as well as the energy intensity of the manufacturing process. Here, it becomes evident that building materials and manufacturing processes are closely linked; if sawmilling practices define how roundwood becomes lumber, machinery will play a key role in shaping the transformation of invasive species into building materials.

An additional consideration shaping the design and construction of *Homegrown* is a desire to create custom panels rather than standard sheets. Custom panels are typically achieved by subtraction from a standard sheet or block of material or by pouring a liquid such as concrete or plaster into a mold or formwork. The construction of custom molds often relies on subtractive processes like CNC milling or foam cutting, with each mold producing only a single unique geometry. This project builds on recent developments in additive manufacturing through deposition, forming, and molding (Kudless, 2011; Oesterle et al., 2012; Vasey et al., 2015; Rusenova et al., 2016; Schumann & Johns, 2019). To construct *Homegrown*, a ground-up machine was developed and entitled "pillow forming." This technique preferences variable over repetitious form, allowing for an infinite number of geometries through a malleable process—the injection and removal of air—repeated over and over again. A series of large-scale inflatable pillows were connected to create an inflatable mold. This process is conceived as a zero-waste, reusable mold for the sustainable construction of lightweight wall panels, built without cost-prohibitive industrial equipment (Schumann & MacDonald, 2021).

Once the computer-controlled pillow-forming apparatus is set in a particular shape, plant material is manually laid up on the mold. Fibers are coated in a liquid bio-based adhesive resin and laid on a flat surface. Material is distributed to match desired thickness, resulting in panels that are alternately thin or thick. Once constructed, *Homegrown*'s four wall panels are assembled to form an exterior room within a walled garden. The project is installed at the Knoxville Museum of Art, between the stepping marble facade designed by Edward Larabee Barnes and the brick elevation of a former factory. The installation measures 10 feet by 10 feet with a rippling interior surface and a flat exterior. The top edge is stepped, suggesting doorways, windows, ledges, and seats (Fig. 10-3).

FIGURE 10-3. *Homegrown*. Knoxville Museum of Art, Knoxville, Tennessee, 2020. (Photo by authors.)

Homegrown repositions the material aesthetics of contemporary architecture, replacing the smooth, crisp, machined tectonics that have defined construction since the introduction of computer-aided design (the Digital Turn) with an architecture that is fuzzy, fluffy, furry, shaggy—in favor of a kind of formal indeterminacy (Carpo, 2013; Johns, 2014; Tabbarah, 2019). Dualities emerge: *Homegrown* is simultaneously primitive and high-tech; it is not permanent, but temporal; the exterior is flat and angular, reflecting conventional sheet materials, while the interior is undulating, suggesting possibilities for customization and sculpted furniture. It is an architecture that requires caretaking and maintenance, like a landscape or an occupant—a rewilding of construction.

Branching Inventory

Branching Inventory (Fig. 10-4) is a structural prototype representing a second approach to working with small-scale invasive plant specimens that are too fine or irregular to be reduced into standardized components (Saslawsky, 2021). If *Homegrown* attempts to aggregate and composite fine material, *Branching Inventory* seeks to catalog and use found material in its whole state, adapting an initial design to an inventory of available parts.

Specifically, the project makes use of fallen branches from the Bradford pear (Fig. 10-5), an invasive and ornamental tree valued for its early flowers and quick growth, qualities that have made it a popular choice for parking lots and new commercial developments. Like many invasive plants, the Bradford pear is still cultivated and sold commercially as a landscaping product. Its wood is valued for small woodworking projects such as bowls and pens turned on a lathe. Nonetheless, it is genetically predisposed to growing acutely angled forks with numerous limbs originating from the same point, in turn making these forks prone to structural failure. The resulting fallen branches and damage to people and property make the Bradford pear a known nuisance.

Branching Inventory is composed of 60 branches from a single Bradford pear tree, each approximately one meter in length. The challenge of using the material at an architectural scale is closely tied to the Bradford pear's characteristics—the wood is of a good quality, but usually occurs in shorter and smaller diameter branches that cannot be feasibly processed into traditional lumber boards. Instead, an approach is developed in which the particular size, arc, and geometry of these irregular pieces can inform the design of a structural assembly.

FIGURE 10-4. *Branching Inventory.* University of Tennessee, Knoxville, Tennessee, 2019. (Photo by authors.)

FIGURE 10-5. Bradford pear specimen. Knoxville, Tennessee, 2019. (Photo by authors.)

First, the design for a structural wall is created using digital modeling software. The wall design consists of an S-curve at the wall's base lofted to a straight line at its top, creating needs for both straight and curved members that range in straightness and curvature.

Maturing digital tools make the next step possible: each branch is 3D scanned using a commercial smartphone app in order to create a digital model inventory of branches. The 3D scans are generated using photogrammetry, a process in which a 3D digital model is created through the analysis of a series of photographs taken at different angles around an object.

Once an inventory of digital branches is compiled, digital modeling software is used to analyze and sort the branches according to certain characteristics: degree of curvature, average circumference, and centerline curve. This allows each individual branch to be located across an undulating structural wall in the location that best matches its characteristics.

Since the branches grow naturally into unique geometries, they will not perfectly match the surface, producing visual deviations. Instead, the original design is reauthored by the material inventory, allowing for the designer to yield authorship and truly collaborate with the material. Finally, the branches are assembled into a physical, structural wall, measuring 30 feet in length and nine feet in height (Fig. 10-6).

FIGURE 10-6. *Branching Inventory*. University of Tennessee, Knoxville, Tennessee, 2019. (Photo by authors.)

As a methodology, *Branching Inventory* avoids the waste produced in the typical milling of roundwood into standard dimensions of lumber. Overall, the use of whole branches reduces energy demands as well as waste during harvesting, production, and on-site construction. This research contributes to a body of work in which found materials inform construction, including wood flitches (Johns & Foley, 2014), tree forks (Devadass et al., 2016; Von Buelow et al., 2018), logs (Larsen & Aagaard, 2019; Zivkovic & Lok, 2020), concrete debris (Clifford & McGee, 2018), bamboo poles (MacDonald et al., 2019), and found objects (MacDonald & Schumann, 2021). *Branching Inventory* is unique within this group in that it uses low-cost, commercially available software and considers how such an approach might expand applications for non-traditional materials—in particular, invasive plant species, which might incentivize the remediation of native ecologies.

Understanding Material

Viewed as a whole, *Homegrown* and *Branching Inventory* beg a questioning of how buildings are designed and produced, the materials with which they are constructed, and how these materials, means, and methods interact with the ecosystem in which a building project is sited. These projects bridge between naturally variable material and design intent, sacrificing some individual design authorship to material inputs to avoid energy consumption and waste. In effect, these demonstrations leverage not just the aesthetic, structural, or behavioral characteristics of grown materials, but also the physical material, and develop a systematic approach to negotiate material intelligence and design intent. The method put forward upends the traditional sequence of design authorship, identifying material capacities in search of an application.

Acknowledgments. The authors would like to thank Kevin Saslawsky and Tyler Sanford, former students at the University of Tennessee, Knoxville, who led the design and fabrication of the *Branching Inventory* project in an architectural design course taught by the authors. Thank you to the Tennessee Invasive Plant Council and Tennessee Citizens for Wilderness Planning for information and assistance in acquiring invasive plant materials. Finally, thank you to the University of Tennessee, Knoxville, College of Architecture and Design for enabling the project *Homegrown* through the framework of the Tennessee Architecture Fellowship, and to the Knoxville Museum of Art for providing an exhibition venue for the work.

Literature Cited

Blaustein, R. J. 2001. "Kudzu's invasion into Southern United States life and culture." Pp. 55–62 *in* J. A. McNeely (editor), *The Great Reshuffling: Human Dimensions of Invasive Species*. IUCN, Gland and Cambridge.

Carpo, M. 2013. "Twenty years of digital design." Pp. 8–14 *in* M. Carpo (editor), *The Digital Turn in Architecture 1992–2012*. John Wiley & Sons Ltd., Chichester.

Clifford, B. & W. McGee. 2018. "Cyclopean cannibalism: A method for recycling rubble." Pp. 404–413 *in* P. Anzalone, M. Del Signore & A. J. Wit (editors), *ACADIA 18: Re/Calibration: On Imprecision and Infidelity*. Proceedings of the 38th Annual Conference of the Association for Computer Aided Design in Architecture. ACADIA.

Cohen, Z., N. Elberfeld, A. Moorman, J. Laucks, S. Kernizan, D. Holmes & S. Tibbits. 2020. "Superjammed: Tunable and morphable spanning structures through granular jamming." *Technology|Architecture + Design* 4(2) 211–220.

Devadass, P., F. Dailami, Z. Mollica & M. Self. 2016. "Robotic fabrication of non-standard material." Pp. 206–213 *in* K. Velikov, S. Ahlquist, M. del Campo & G. Thün (editors), *ACADIA 2016: Posthuman Frontiers: Data, Designers, and Cognitive Machines*. Proceedings of the 36th Annual Conference of the Association for Computer Aided Design in Architecture. ACADIA.

Finch, B. 2015. "The true story of kudzu, the vine that never truly ate the South." *Smithsonian Magazine*. https://www.smithsonianmag.com/science-nature/true-story-kudzu-vine-ate-south-180956325/

Harron, P., O. Joshi, C. B. Edgar, S. Paudel & A. Adhikari. 2020. "Predicting kudzu (*Pueraria montana*) spread and its economic impacts in timber industry: A case study from Oklahoma." *PLoS One* 15(3):e0229835. https://doi.org/10.1371/journal.pone.0229835

Johns, R. L. 2014. "Augmented materiality: Modelling with material indeterminacy." Pp. 216–223 *in* F. Gramazio, M. Kohler & S. Langenberg (editors), *Fabricate 2014: Negotiating Design and Making*. UCL Press, London.

Johns, R. L. & N. Foley. 2014. "Bandsawn bands: Feature-based design and fabrication of nested freeform surfaces in wood." Pp. 17–32 *in* W. McGee & M. Ponce de Leon (editors), *Robotic Fabrication in Architecture, Art and Design 2014*. Springer, Cham.

Kudless, A. 2011. "Bodies in formation: The material evolution of flexible formworks." Pp. 98–105 *in* J. S. Johnson, B. Kolarevic, V. Parlac & J. M. Taron (editors), *ACADIA 2011: Integration through Computation*. Proceedings of the 31st Annual Conference of the Association for Computer Aided Design in Architecture. ACADIA.

Larsen, N. M. & A. K. Aagaard. 2019. "Exploring natural wood: A workflow for using non-uniform sawlogs in digital design and fabrication." Pp. 500–509 *in* K. Bieg, D. Briscoe & C. Odom (editors), *ACADIA 19: Ubiquity and Autonomy*. Proceedings of the 39th Annual Conference of the Association for Computer Aided Design in Architecture. ACADIA.

MacDonald, K. & K. Schumann. 2021. "Twinned assemblage: Curating and distilling digital doppelgangers." Pp. 693–702 *in* A. Globa, J. van Ameijde, A. Fingrut, N. Kim & T. T. S. Lo (editors), *Projections: Proceedings of the 26th International Conference of the Association for Computer-Aided Architectural Design Research in Asia*. CAADRIA, Hong Kong.

MacDonald, K., K. Schumann & J. Hauptman. 2019. "Digital fabrication of standardless materials." Pp. 266–275 *in* K. Bieg, D. Briscoe & C. Odom (editors), *ACADIA 2019: Ubiquity and Autonomy*. Proceedings of the 39th Annual Conference of the Association for Computer Aided Design in Architecture. ACADIA.

Oesterle, S., A. Vansteenkiste, A. Mirjan, J. Willmann, F. Gramazio & M. Kohler. 2012. "Zero waste free-form formwork." Pp. 258–267 in J. Orr, M. Evernden, A. Darby & T. Ibell (editors), *Proceedings of the Second International Conference on Flexible Formwork*. BRE CICM, University of Bath, Claverton Down.

Rusenova, G., K. Dierichs, E. Baharlou & A. Menges. 2016. "Feedback- and data-driven design for aggregate architectures: Analyses of data collections for physical and numerical prototypes of designed granular materials." Pp. 62–72 in K. Velikov, S. Ahlquist, M. del Campo & G. Thün (editors), *ACADIA 2016: Posthuman Frontiers: Data, Designers, and Cognitive Machines*. Proceedings of the 36th Annual Conference of the Association for Computer Aided Design in Architecture. ACADIA.

Saslawsky, K., T. Sanford, K. MacDonald & K. Schumann. 2021. "Branching Inventory: Democratized fabrication of available stock." Pp. 513–522 *in* A. Globa, J. van Ameijde, A. Fingrut, N. Kim & T. T. S. Lo (editors), *Projections: Proceedings of the 26th International Conference of the Association for Computer-Aided Architectural Design Research in Asia*. CAADRIA, Hong Kong.

Schumann, K. & R. L. Johns. 2019. "Airforming: Adaptive robotic molding of freeform surfaces through incremental heat and variable pressure." Pp. 33–42 *in* M. H. Haeusler, M. A. Schnabel & T. Fukuda (editors), *Intelligent & Informed: Proceedings of the 24th International Conference of the Association for Computer-Aided Architectural Design Research in Asia (CAADRIA 2019)*, Vol. 1. CAADRIA, Hong Kong.

Schumann, K. & K. MacDonald. 2021. "Pillow forming: Digital fabrication of complex surfaces through actuated modular pneumatics." Pp. 292–301 *in* K. Dörfler, S. Parascho, J. Scott, B. Bogosian, B. Farahi, J. L. García del Castillo y López, J. A. Grant & V. A. A. Noel (editors), *ACADIA 2021: Realignments: Toward Critical Computation*. Proceedings of the 2021 Conference of the Association for Computer Aided Design in Architecture. ACADIA.

Tabbarah, F. 2019. "Where do the twigs go?" Pp. 152–161 *in* K. Bieg, D. Briscoe & C. Odom (editors), *ACADIA 2019: Ubiquity and Autonomy*. Project Catalog of the 39th Annual Conference of the Association for Computer Aided Design in Architecture. ACADIA.

Vasey, L., E. Baharlou, M. Dörstelmann, V. Koslowski, M. Prado, G. Schieber, A. Menges & J. Knippers. 2015. "Behavioral design and adaptive robotic fabrication of a fiber composite compression shell with pneumatic formwork." Pp. 297–309 *in* C. Perry & L. Combs (editors), *ACADIA 2015: Computational Ecologies: Design in the Anthropocene*. Proceedings of the 35th Annual Conference of the Association for Computer Aided Design in Architecture. ACADIA.

Von Buelow, P., O. Oliyan Torghabehi, S. Mankouche & K. Vliet. 2018. "Combining parametric form generation and design exploration to produce a wooden reticulated shell using natural tree crotches." *In* C. Mueller & S. Adriaenssens (editors), *Proceedings of the International Association for Shell and Spatial Structures (IASS) Symposium 2018: Creativity in Structural Design* 2018(20): 1–8.

Zivkovic, S. & L. Lok. 2020. "Making form work: Experiments along the grain of concrete and timber." Pp. 116–123 *in* J. Burry, J. Sabin, B. Sheil & M. Skavara (editors), *Fabricate 2020: Making Resilient Architecture*. UCL Press, London.

CHAPTER 11

An Ethic of Care in Basketry: Weaving with Invasive Vines

Katie Grove

I am standing in a circle of women on the grass of an overgrown dirt road in the woods. The midmorning sun is slowly ascending and the dew has just evaporated from the lush undergrowth, leaving a feeling of hazy oncoming heat in the air. It's July and rich, abundant, unstoppable life is all around us. Wineberries are glistening in the dappled sunshine. The voices of robins, invigorated by the morning reprieve from the deep summer heat, occasionally sing out from the forest. At the edges of the clearing, leaves belonging to a hodgepodge of trees, flowers, shrubs, and vines are stretching out toward open space, sunshine, and the prospect of further growth.

The women in the group are here to learn how to weave baskets with wild foraged plant materials in a nine-month-long workshop series that I've taught for six years in the woodlands of New York State. Foraging brought us to this overgrown woodlot; we are here for weaving materials. Backpacks, loppers, and pruning shears in their woven sheaths rest on the ground at our feet—the essential tools of the forager. It is the second year that this particular group has spent together, evidenced by the comfortable laughter and chatting as they turn to each other to check in about life since the last time we met. Through the shuffling of feet and morning greetings I call everyone's attention to the tasks of the day. As always, we start with a check in, going around in a circle and sharing something we are excited about or grateful for. In this way each person has space to talk about what is alive for them in this moment. It brings us together as a group and allows us to arrive, shaking off the stresses and to-do lists of the rest of our lives as we come into a space of learning and connection. As the circle returns to me I echo so much of what the others expressed: gratitude to be here, gratitude for the plants we are harvesting, and gratitude for the land.

After a few breaths to settle us in I begin. "Alright everyone—you can stand or sit. There is something I want to read to you. And then I want to hear what you think." I pull out a sheet of paper and begin to read a passage aloud. It's from a treatise on Ainu ethnobiology by Dai Williams, published in 2017, but I omit this information along with the name of the plant the passage describes.

> The branches are gathered anywhere from December to March, the bark is split vertically, and the inner wood core removed and discarded . . . The bast layer is then carefully separated from the bark, bundled, and soaked in hot water for a few minutes, which causes the surface to turn deep green. It is next bleached on the surface of the snow for two weeks, by which time it turns pure white. If there is no snow, it is hung from a horizontal pole for about a month . . . The resulting white fibre is called **retahai**, *reta* meaning white . . . As it is extremely strong, the Ainu learned it was an ideal material for making their tumplines, or **okae-tara**, for cords, rope, nets, bow strings, sword straps, and a belt called **ra-un-kut** . . . The **ra-un-kut** is 3.6 m long; each warp is made up of eight extremely finely spun threads. It was woven by her mother or maternal grandmother and given to a menstruating girl as a confirmation of her matrilineage, with woven designs representing this. The Ainu believed that, without the belt, a woman would not be able to find her place in the afterlife. It was worn beneath her clothes under the bosom. Once her mother had placed the belt, nobody but the young woman or her husband was allowed to touch it . . . (Williams, 2017).

As my voice trails off I let the lazy robins tutting and the distant sound of a summer tanager fill the space. "What plant do you think they're describing?" The others pause and think, closing their eyes and looking back over the past year and a half we've spent together, learning about a myriad of different plants. They pick up on the part of the plant described—the bast, or inner bark. "Basswood?" someone asks, referring to a tree we processed together the previous summer for its beautiful and strong bast fiber. I shake my head and wait.

"Elm?" Another calls out, naming a native tree whose bark we've used for various projects. "Nope." I answer with a smile. Some other guesses ring out—nettle, dogbane, milkweed. They suggest various plants we have worked with or that they have heard of that provide bast fiber for weaving.

Their ideas exhausted, I gesture dramatically with my arms to the tangle of vines covering the undergrowth around us and stretching up into the trees. "Bittersweet!" The revelation is a surprise, met with big smiles, shaking heads, and many questions.

Bittersweet, or Celastrus orbiculatus, is a vine that everyone in this group knows well. It is known primarily as an enemy to gardeners, environmentalists, and anyone who has ever bought a house in the northeast and sought to tame the unruly wood edge. It is a vigorous and enthusiastic grower that, like many vining plants, makes its way in the world by covering everything in its path, twisting around trees and shrubs and eventually strangling them as it climbs. In the fall its bright red and orange berries attract birds that readily eat and spread the seeds, resulting in tiny new vines sprouting up even in shady places. If a person trying to rid their woods of bittersweet were asked to describe this plant in one word it would inevitably be this: invasive. As defined by the U.S. Forest Service (2024) an invasive plant is one that is both "non-native (or alien) to the ecosystem under consideration" and "whose introduction causes or is likely to cause economic or environmental harm or harm to human health." According to this perspective bittersweet fits the bill, being native to Southeast Asia and having the propensity to grow unchecked. It also happens to be a wonderful material to make random weave, ribbed, and many other styles of basket (Fig. 11-1).

Slowly the group's attention returns from letting the news settle in that a plant maligned in America could historically be a culturally valuable species to another group, the Ainu, an indigenous people on the island of Hokkaido in Japan. From here questions and thoughts pile on as the group dives into a discussion on not only bittersweet, but many other invasive vines used in previous basket classes. Wisteria, kudzu, honeysuckle vine, and akebia all are brought up. What does it mean to be invasive? Can a native plant have invasive qualities? Are invasive plants beneficial in some ways? What is the best and most effective way to remove them? Is it our responsibility to do this as humans who are continually altering ecosystems? The conversation is rich and complicated—there is no consensus. It's as tangled and knotty as the wall of bittersweet we are here to harvest.

FIGURE 11-1. Wild Basketry student holding a random weave basket made of bittersweet (*Celastrus orbiculatus*). (Photo by author.)

Untangling the Thicket

The subject of invasive plants is one that I think about a lot as a basketmaker and a person who cares deeply about the land that I call home along the creeks and hills of the Rondout Valley in New York State. Just like my students' conversation in the woods, my thoughts around these opportunistic, zealous, insuppressible plants are fraught with contradictions. They are further convoluted by my position as a descendant of western European settlers and immigrants. My own ancestors neither tended the forest for generations, nor worked with these particular plants in their own homelands. While I respect bittersweet's natural tendency to grow and thrive, I also seek to nurture diversity in the woods around my home where it overtakes native shrubs. I am thrilled when a local landowner asks me if I want to bring my class to harvest materials and I see healthy, thriving vines. I also feel sadness when I find bittersweet girdling the trunk of a basswood tree and I experience an overwhelming urge to "save" the tree by spending hours cutting vines. I feel disappointed when I see places where kudzu has been sprayed by pesticides. After many years of seeking out and harvesting invasive vines to weave into baskets one thing has become clear to me: invasive plants have

Building Relationship with the Land

The woods and fields around us are alive with multifarious types of invasive plants that are excellent for basketry, from phragmites leaves to the barks of certain trees. However, it is the vines that I focus on most in my classes. Bittersweet is not the only vine capable of being woven. I let my mind dance around the names of those found most commonly in my stomping grounds on the East Coast of the United States—*Wisteria floribunda* (Japanese wisteria, Fig. 11-2), *Lonicera japonica* (honeysuckle vine), *Akebia quinata* (chocolate vine), *Ampelopsis brevipedunculata* (porcelain berry), and perhaps most of all, "the vine that ate the South" *Pueraria montana* var. *lobata*, or kudzu. To differing degrees these vines are all vigorous growers categorized as invasive. Not only this, but they are typically conspicuous—people tend to notice the sinuous forms of bittersweet encircling the trunks

FIGURE 11-2. A woodland overtaken by wisteria (*Wisteria floribunda*). All of the plants visible on the forest floor are wisteria ground runners. (Photo by author.)

of trees, or a carpet of pure wisteria in the open woods. Vines are excellent ambassadors for basketry and a good place to launch one's journey with wild basketry, which essentially begins with learning to identify plants. Teaching students to recognize common plants that they are most likely to actually encounter makes it much easier for those students to begin parsing apart "the wall of green" that so many people new to working with nature tend to see when they walk into the woods.

❖ ❖ ❖

The prolific nature of these vines, plus the lack of natural controls in the forms of insects, animals, and human intervention may lead them to overtake other vegetation, but the vines are not to blame for this. They are simply taking advantage of the opportunity to do what all plants do: grow, reproduce, and thrive, especially in the disturbed ecosystems that are a byproduct of human activity. However, when a friend reaches out, asking if I want to harvest the wisteria growing wildly in their woodlands, I see an opportunity not only to harvest weaving material, but to remove the vines which are choking trees and other vegetation.

Unfortunately, invasive vines like wisteria are very difficult to permanently remove—they can resprout from root pieces and their seeds can stay alive in the soil for years. Typical methods include mowing, manually removing vines and roots, and applying pesticides. The application of pesticides is typically recommended, even after manual removal, which renders the environment unsuitable for foraging weaving material. For those who don't want to use pesticides: "cutting alone is only effective at controlling the [bittersweet] vines when resprouts are repeatedly cut until the root system is exhausted. This will take multiple cuttings annually over several growing seasons" (Templeton et al., 2020).

While returning to a spot annually to remove vines without pesticides is inconvenient to many, to the basketmaker this provides an opportunity to build a relationship with the land and all the plants growing there—including the invasive ones. In my own practice, by virtue of spending time visiting a spot year after year, I get to know the types of trees present there, observe the growth habits of the vines overtaking them, and develop my awareness as I watch what happens over time. Are the trees recovering? Are any other species returning, or is it just more bittersweet coming up along the ground? As I begin to truly know and care for a place I find myself motivated not only to gather weaving material, but to steward the land to the best of my ability.

FIGURE 11-3. Coils of wisteria (*Wisteria floribunda*) removed from a woodland in New York. (Photo by author.)

Removing wisteria becomes not only about gathering useful vines, but also about encouraging the survival of the trees they are strangling. At the same time I come to appreciate wisteria for its tenacity and ability to grow such wonderful weaving material (Fig. 11-3). It is difficult to truly understand the scope of consequences over time that human interventions have in an ecosystem. However, it is my goal and hope that by harvesting invasive vines year after year I can help move the land toward balance in the ecosystem, where a greater diversity of plants have a chance to thrive.

Expressing Gratitude with Equality

Alongside learning how to identify, forage, store, prepare, and finally weave with natural materials my students and I also practice treating the plants we harvest with the same respect and care that we would for a friend. This may look like asking a tree for permission to be harvested—and then using both our scientific brains and our intuitive hearts to look and listen for an answer. We draw on important and age-old questions advocated for in Robin Wall Kimmerer's essential book, *Braiding Sweetgrass*, as we harvest. Is it the only one of its kind growing in the area? Is it struggling to grow against its neighbor? Is it about to go into flower, wherefore harvesting it will not allow the plant to go to seed for the season? Is there a feeling deep in the gut that this plant is better off where it is? All of these clues and more combine to help us make decisions about harvesting with respect (Kimmerer, 2013).

The practice of expressing gratitude, asking permission, and engaging with plants in a way that acknowledges their aliveness can be a brand new experience to many people. It might be easier to step into when harvesting a large and autonomous tree, or a precious native flower. However, this procedure of care is often forgotten when one is dealing with invasive plants. I reject the ethic that the best part about making things with invasive plants is that you can harvest them indiscriminately without guilt. We may have the ultimate goal of removing wisteria from an area, but there are many paths to get to that goal. In the end I maintain that what we do matters—*but how we do it matters more*. Yes, stomping into a patch of wisteria and yanking vines out of a tree may result in removal of the vines. It also results in ripped vines, the branches of the tree being damaged, and wildflowers being crushed underfoot. By approaching harvest as a slow practice and expressing gratitude for the vines in the same way we would for a treasured native species, it gives foragers the opportunity to embody open mindedness and kindness—qualities that can serve us well in any aspect of life.

Honoring the Cultural Histories of Invasive Plants

When thinking about the potential value of a vine like wisteria or kudzu, which is often scorned in America, it helps to remember that this plant *comes from somewhere*. Looking at the natural and cultural history of an invasive plant in its homeland reminds us that this plant is native somewhere, and may be considered completely differently there. In those places they fill ecological niches, are deeply entwined with human history, are called by local names, and may even be celebrated. This knowledge can serve as a launching point for learning about plants previously just thrown in the weed pile. While engaging in this investigation, considering one's own race and ancestry is fundamental. A person of Ainu descent will have a completely different way of relating to the cultural history of bittersweet than one of European descent. For someone of the latter group, like myself, it is important to continually examine if your practices and approaches are engaging in cultural appropriation. Gaining knowledge of bittersweet's value and uses by the Ainu in Hokkaido helps to open the heart toward valuing it; it is not an invitation to appropriate their methods and practices. Can the vine in question be used in a new and different way? Are there styles of basketry it can be incorporated into that are global in origin, or unique creations? These plants have so much potential and experimenting with them will continue to yield new ways of working.

Bittersweet is not the only invasive vine valued in its place of origin. Take akebia, kudzu, wisteria, honeysuckle, and several other vines used for basketry and trace them from the eastern part of the United States back to their points of origin and you will find yourself nearly 7,000 miles away in the temperate forests of East Asia. Both New York State and much of the archipelago of Japan, for example, share a similar climate and four seasons, with hot humid summers and snowy winters. Each location houses native representatives from notable genera of basketry plant groups such as *Ulmus* (American elm and Manchurian elm) and *Tilia* (basswood and Japanese linden), as well as countless other traditionally used native species, including cattails and sedges. It's not surprising at all that vines such as kudzu and bittersweet evolving in the forests of Japan would find themselves right at home in the mixed temperate deciduous forests of the United States.

Kudzu is perhaps the best example of a vine that, despite being maligned in the United States (particularly the South), has the potential to be a resource worthy of our respect and care.

As a weaving material kudzu has many superpowers. Its ground runners can grow incredibly long and when split soon after harvesting make excellent weavers in ribbed, woven, and twined baskets. The thicker diameters are strong and dynamic—perfect for hoops, sculptural baskets, and larger work. Much like bittersweet, the bark can be peeled from the vine and either woven as is or processed down into fiber. I get excited when I drive down south and see an oppressive wall of dead leaves in the winter that marks an excellent kudzu gathering spot; knowing that beneath it lie ground runner vines that can stretch for 30 or 40 feet. A holy grail of weaving material. However, my initial excitement is inevitably dashed when I remember I am driving on a busy road and this kudzu has likely been sprayed with pesticides, rendering it unsafe to forage. Knowing kudzu's potential as a resource I wonder if pesticide application could be reduced in favor of more holistic methods of control (such as annual visits by basketmakers).

Looking at kudzu's natural and cultural history in its native lands, particularly in Japan, reveals a long and storied background. It has been harvested and processed for food, medicine, and fiber since the Jomon, or Neolithic, era (Dusenbury, 1992; Wong et al., 2011; Wang, et al., 2021). As a fiber plant kudzu, or kuzu as it is called in Japan, has a history that is rich and textured, being a prized material for its lustrous fibers, revealed only after a slow and many-stepped process. Besides being harvested as a fiber-producing plant for weaving textiles, kudzu, as well as wisteria, bittersweet, and others, were used to make all manner of intense utility and cultural value, including rope, baskets, nets, and clothing (Cort, 1989: 378). Despite the introduction of hemp, ramie, and later flax, the use of kudzu fiber has persisted in Japan, due to both its symbolic value in ritual contexts and usefulness for clothing in more remote areas. This and the revival of traditional crafts as a cultural movement have tenuously allowed kudzu fiber usage to continue as a tradition despite modernization and the availability of fiber crops that are easier and faster to utilize (Cort, 1989: 407).

What does it change to know that kudzu, a plant typically doused with pesticides and removed at all costs, is a culturally valuable plant, as well as an economic commodity in its homeland? I do not suggest we engage in cultural appropriation of the ceremonial and traditional clothing, textiles, and baskets made with kudzu in Japan and mimic the ways that it has been used in its native lands. This would be an injustice and disrespectful to the traditional cultures that value it. What I do suggest is that we re-examine our own relationship with kudzu and start to ask questions about how we can utilize this plant in a respectful way.

Gathering Together

After spending several hours in the woods gathering bittersweet we return back to my studio, unloading coils and unruly piles of vines from truck beds. There, we sort them into different diameters and set them alongside coils of wisteria, akebia, and split kudzu already drying in the studio since their harvest over the previous year. The fresh vines don't need soaking, but the dried ones do. We put them into pots of steaming water to heat until they regain flexibility and can be woven. After a break for lunch we finish sorting the vines for the different parts of our basket, a three-hooped ribbed basket, also known as a hen basket for its similarity in shape to a sitting hen. The ribbed basket has multiple parts—the hoops, which must be sturdy, the ribs, which need to be flexible yet still strong, and the weavers, which must be able to bend readily and fit in between each rib to create a snug weave. For each of these purposes we utilize a different material (Fig. 11-4).

FIGURE 11-4. A ribbed basket woven by the author in bittersweet (*Celastrus orbiculatus*), wisteria (*Wisteria floribunda*), akebia (*Akebia quinata*), and kudzu (*Pueraria montana* var. *lobata*). (Photo by author.)

Weaving a basket with vines can be a wild process. I describe it as "wrestling alligators" to students who are having difficulty manipulating thick lengths of wisteria that defy shaping. "Close your eyes and begin to bend the vine—think of it like you are a massage therapist sensing all those woody fibers running throughout the vine. You want to be strong and confident, but also gentle. If you try to bend it too quickly it will snap. But if you move carefully, listening to the vine, you can bend it in half."

The resounding snap of a vine cracking fills the air along with laughter and frustrated sighs as students practice bending the vines into hoops, forming the frame of our ribbed baskets. As a basketmaker you have to read the material and each has its own strengths and weaknesses. Bittersweet is not very flexible, but it is strong, making it great for the hoops and ribs. As a weaver, which must be flexible and long, it is not ideal. On the other hand, split kudzu or thin akebia vine make amazing weavers. With time and patience, students begin to finalize

FIGURE 11-5. The author weaving a ribbed basket during a workshop. (Photo by Stefan Lisowski.)

the hoops and ribs, which make up the skeleton of this style of basket. We then choose the thinnest akebia vines to begin weaving the handle, where they must fit into tiny spaces in order to create a tight weave. Once the first couple of inches are woven and the ribs are secured (they have the propensity to fly out of the basket, inciting what I affectionately call "rib rage") then the real weaving starts. Over and under, over and under.

The struggle becomes a meditation and we settle into a rhythm of weaving, splicing in new materials, chatting, and sometimes just sitting in silence and letting the quiet, gentle power of making in community flow around us like rocks in a river (Fig. 11-5). Vines that are typically considered at best a nuisance and at worst a villain become our muses. We discuss the intricacies of each one and folks brainstorm where they can find their own sources of vines—places where they can forage valuable material while also helping out local landowners and the ecosystem. In this practice we can seek balance, finding a way to be grateful for even the most maligned of plants while working toward more plant diversity in an ecosystem.

The Slow Practice of Weaving a Future

It is my greatest desire that anyone who comes to my basketry workshops leaves with the inspiration to not only look at their landscape with new eyes, but to start developing a reciprocal relationship with it. Just like weaving a basket, this is a slow practice, requiring time, patience, and care. As we work my students trace their fingers over different textures of plant material woven into their baskets and engage in conversation around invasive species and the consequences of our actions as humans on this earth. Through the sharing of vulnerable thoughts, difficult questions, and personal stories our collective knowledge grows like a wellspring. Perhaps more than any other type of plant the sight of a length of kudzu running along the ground into the thicket or bittersweet twisting around a sapling has the potential to represent a shift in perspective. Can the woods become not just a place to ignore or take from, but a place to steward and care for? And can the invasive plants growing there become a resource worthy of our time, care, respect, and attention as well? While I watch my students and friends sit in community, coaxing strands of vine over and under, I know that the answer is yes.

Literature Cited

Cort, L. 1989. "The changing fortunes of three archaic Japanese textiles." Pp. 377–415 in A. B. Weiner & J. Schneider (editors), *Cloth and the Human Experience*. Wenner-Gren Foundation for Anthropological Research, Smithsonian Institution Press, Washington.

Dusenbury, M. 1992. "A wisteria grain bag and other tree bast fiber textiles of Japan." Pp. 259–270 in *Textiles in Daily Life. Proceedings of the Third Biennial Symposium of the Textile Society of America, September 24–26, 1992*.

Kimmerer, R. W. 2013. *Braiding Sweetgrass. Indigenous Wisdom, Scientific Knowledge and the Teachings of Plants*. Milkweed Editions, Minneapolis.

Templeton, S, A. Gover, D. Jackson & S. Wurzbacher. 2020. "Oriental bittersweet." Penn State Extension. https://extension.psu.edu/oriental-bittersweet

U.S. Forest Service. 2024. "Invasive plants." U.S Department of Agriculture. https://www.fs.usda.gov/wildflowers/invasives/index.shtml

Wang, Z., K. Deng & Z. Peng. 2021. "The origin and development of Chinese warp twisted fabric." *Asian Social Science* 17(2): 90–98. https://doi.org/10.5539/ass.v17n2p90

Williams, D. 2017. *Ainu Ethnobiology*. Contributions in Ethnobiology Monographic Series. Society of Ethnobiology, Tacoma.

Wong, K. H., G. Q. Li, M. L. Kong, V. Razmovski-Naumovski & K. Chan. 2011. "Kudzu root: Traditional uses and potential medicinal benefits in diabetes and cardiovascular diseases." *Journal of Ethnopharmacology* 134: 584–607. https://doi.org/10.1016/j.jep.2011.02.001

CHAPTER 12

Invasive Color: Using Invasive Species as Natural Dyes

Theresa Hornstein

Humans have a long history of using natural dyes. Barber (1991) references dyed fibers found in neolithic archeological sites across Anatolia and Mesopotamia. Vajanto (2014) has reported on Viking-era dyes. While many of the dyeing processes have been lost to history as aniline dyes replaced natural in the 1850s, several of the plants now considered to be invasive species—tansy, buckthorn, and teasel, for example—have long histories as dye sources. Other invasive species have not been documented in the literature, but still provide beautiful color. The basic dye processes described here can act as a guide for experimentation.

Materials and Methods

Natural dyes work best with natural fibers. Animal fibers include wool and silk. These fibers are proteins and readily absorb most natural dyes. Silk and different types of wool take up the dyes slightly differently. The examples in this chapter all use merino wool. Alum is the most common mordant to use with protein fibers. Cellulose is a material produced by plants. Linen, cotton, and bamboo are examples of cellulose fibers. These are carbohydrates and do not pick up the dyes as strongly or may pick up different dye components than protein fibers. Tannins are used with cotton to increase the dye uptake. Some people will soak cellulose fibers in soy milk and allow it to dry before dyeing.

Not all colors obtained from natural dyes are permanent, or fast. While most colors will eventually fade, there are some which fade or change so quickly as to be fairly useless for dyework that is exposed to light or washing. These dyes are referred to as fugitive. Mordants are used to improve the fastness of dyes, acting as a bridge, joining the dye molecules to the fiber. Mordants can also act as modifiers, chemicals that shift or alter the color a dye produces (Mairet, 1916). Mordants are either used to pretreat the fiber before dyeing or added to the dye bath during the dye process. Modifiers can be added to the dye bath or used as a dip after dyeing. Using modifiers and mordants can dramatically expand the range of colors produced from a single dye.

The most common mordant throughout history is the aluminum salt, alum, in the form of aluminum sulfate, aluminum potassium sulfate, or aluminum ammonium sulfate. This is combined with cream of tartar (potassium bitartrate), increasing the absorption of the alum by the wool (Dean, 1999). Silk, linen, and cotton do not require cream of tartar, but do show greater fastness if alum is used. The standard process to mordant wool with alum is to create a solution containing 5%–8% alum and 4%–7% cream of tartar relative to the weight of the dry fiber. Dissolve the alum/cream of tartar in enough warm water to allow the fiber to move freely. Add the fiber and allow to soak for an hour. Squeeze out the excess solution and proceed to the dye bath. Premordanted fiber can also be hung to dry and dyed later. Rewet the fiber before dyeing.

Iron, usually in the form of iron oxide (rust) or iron sulfate, acts as both a mordant and a modifier. It adds a darker tone to the fibers, referred to as saddening. Using iron can shift browns to grays or blacks. Yellows usually shift to greens. However, large amounts of iron can weaken fibers.

Copper, usually in the form of blue/green copper sulfate, acts as both a mordant and a modifier. It enhances the greens, shifting yellow-green fibers to a richer green. Steeping copper pipe or old pennies in vinegar produces a blue liquid which can be used to soak the fiber before dyeing or added to the dye bath.

Altering the pH of the dye bath or treating the fibers after dyeing can alter the color of some dyes (Fig. 12-1). The pH scale runs from 0 to 14, where 7 is neutral. The further below 7 the pH is, the more acidic something becomes. The further above 7, the more basic, or alkaline, it becomes. Either extreme can be damaging to both skin and fibers.

Vinegar is a common acid added to lower the pH. Citric acid (sour salt), fermented fruits, citrus juice, and ascorbic acid (vitamin C, a color-keeper used in canning) can also be used to lower the pH. Baking soda (sodium bicarbonate), washing soda (sodium carbonate), and household ammonia (dilute ammonium hydroxide) are common bases used. Many old dye books use chalk (calcium carbonate) as a base. The effects vary depending on the specific dye, the mordant used, and how much the pH is altered.

One of the most variable components in natural dyeing is the water used. To produce consistent results, use distilled water. If using city or well water, there may be a different pH and trace minerals that can alter the results.

Natural dyes can vary from year to year and location to location. The soil minerals, the time of year, even temperature can alter the colors produced. Different fibers give different results, even in the same dye pot. Despite best attempts, it is difficult to get exactly the same color from one dye session to another.

Basic Safety Measures

Just because something is natural does not mean it is harmless. Some plants also produce irritants, triggering allergic reactions or contact dermatitis. Others produce unpleasant effects if ingested. Arsenic, lead, and chrome—all toxic—have been used as mordants in the past. Concerns about safety need to be taken into account while dyeing. Most dyers work on a small scale in the kitchen. These simple rules will help to make the dyeing experience a safe one.

1. Keep the dye equipment separate from equipment that will come in contact with food. Use different cutting boards, measuring tools, pots, and stirrers. Wash up the dye equipment with its own dedicated sponge. When finished, store the dye equipment in a dedicated and clearly labeled box away from the kitchen.

2. Choose the materials wisely. Make sure the plant material, mordants, and modifiers are safe. Irritants like leafy spurge or wild parsnip should not be used as dyes.

3. Work in a well-ventilated area. Heating the dyes can release unpleasant fumes.

4. Waterproof gloves are a good idea both to minimize direct contact with the dye material and to limit dye transfer to the skin.

Extracting the Dyes

Producing dye baths is the easiest method for dyeing fiber. The first step is to get the dye out of the source material. Most materials require an equal weight of dye material to fiber. Less plant produces paler colors. There are a variety of methods that work with different materials. Water is the common liquid used, but experiment. Some pigments are soluble in alcohol rather than water. If using alcohol, be careful when heating the dye bath.

Soft materials can be chopped or torn into small bits before processing. This increases the surface area and helps the colors leach out into the dye bath. This works well with leaves and flower parts. Harder materials like wood or bark can be grated, chipped, or shaved. Barks also benefit from soaking before processing. Most barks contain large amounts of tannins which are released when boiled. Tannins add a brown color to the dyes. To avoid tannins darkening the dye, keep the temperatures lower when extracting the dyes.

Soaking dyestuffs in hot water is one of the oldest dye methods and the one most people are familiar with. The dye material is heated in water, allowed to steep, and the dye liquid strained off for use. The fibers are then added to the bath, and they absorb the color. Typically, the bath is heated and held at a simmer for an hour or more and the fiber allowed to cool in the bath to produce deeper, more permanent results. After dyeing, rinse the fiber in cool water until the rinse water runs clear.

Exhaust baths reuse the dye bath after the initial dyeing. Introducing more fiber into the "exhausted" bath will produce lighter shades. This is good for producing a variety of shades in a single color or for combining dyes. In Figure 12-2, the original dyeing pulled browns from the buckthorn bark bath. The original bath was produced by soaking buckthorn bark in equal parts household ammonia and water. The pH was 10. After soaking overnight the dye was strained. Alum-mordanted fiber was added and the bath heated. The subsequent exhaust baths used the same kind of fiber and show more of the pink produced by buckthorn bark under alkaline conditions.

Heat helps release the pigments and fix them to the fibers. Don't allow the dye bath to get too low. The fibers need to keep moving. When working with wool, avoid abrupt temperature changes, agitation, and hard boiling. This can felt the fibers.

Another method is to solar dye. The dye and fiber are placed in a jar or plastic bag and left in the sun for a day or more. The sun provides the heat.

Solar dyeing takes longer, but minimizes the chance of felting if working with wool and requires little tending. Just set it up and walk away.

Over dyeing is a process where a fiber is dyed one color then dyed again with a second color. Using this technique can dramatically expand the range of colors produced from a relatively few basic dyes.

Dye Examples
Common Buckthorn (*Rhamnus cathartica*) & Glossy Buckthorn (*Frangula alnus*)

Both common and glossy buckthorn are classic invasive species, crowding out native plants, creating nearly impenetrable thorny thickets, and disrupting ecosystems. For all the problems buckthorn creates, it is one of the most versatile dye plants. The bark, berries, and leaves all produce dyes.

The bark separates most easily in the spring and early summer, but buckthorn bark can be gathered any time of year. Use a sharp knife to strip the bark from the tree. Freshly peeled bark appears yellow to deep orange (Fig. 12-3). It rapidly oxidizes and turns nearly black. This does not alter the dye produced. It can be used immediately or dried for later use.

FIGURE 12-1. Buckthorn berry dye on alum-mordanted wool. **Upper:** pH 10, producing a fast green. **Lower:** same wool in a pH 3 dye bath, producing a fugitive pink. (Photo by author.)

FIGURE 12-2. Exhaust baths produce lighter versions of the original bath and may allow paler undertones to show through. Buckthorn bark dye in ammonia and water on wool: **left**, original dye bath; **middle**, first exhaust bath; **right**, second exhaust bath. (Photo by author.)

FIGURE 12-3. Freshly cut buckthorn bark showing the classic yellow/orange inner bark. (Photo by author.)

FIGURE 12-4. Wool yarns simmered with buckthorn bark. **Top to bottom:** no mordant; with alum, pH 7; with alum, pH 4; with alum, pH 10; with alum and iron. (Photo by author.)

Buckthorn bark can be processed two different ways. Gently simmering the bark in water produces golds and browns (Fig. 12-4). Increasing the heat releases more tannins and produces a deeper gold to brown color depending on the age of the bark. Older bark produces deeper colors. Twigs produce lighter golds and yellows.

Buckthorn bark can also be soaked in a solution of base and water. Strain the dye and add enough water to allow the fiber to move freely. Be sure to check the pH. The solution needs to remain basic. A pH between 9 and 11 produced the best color. Add the fiber and barely simmer. Allow the fiber to cool in the pot. This produces a rust red dye, almost fox color (Fig. 12-2).

Fresh buckthorn leaves produce yellows to slightly green yellows (Figs. 12-5, 12-6). Buckthorn holds its leaves well into the winter. One easy way to identify a stand of buckthorn is to look for the still-green leaves into December, long after all the other trees have dropped their leaves. This is one of the few yellows that does not shift green with the addition of an iron modifier.

Fresh buckthorn berries, gathered in the fall and early winter, produce vibrant greens that are fast and a fugitive pink (Fig. 12-7). The berries start green and ripen to a deep blue/black. The riper berries give the best colors. To create the dye bath, simmer the berries in water. Filter the dye carefully. Add the fiber to the simmer for at least an hour.

INVASIVE COLOR 247

FIGURE 12-5. Buckthorn leaves need to be used fresh. (Photo by author.)

FIGURE 12-6. Buckthorn leaf on wool. **Top to bottom:** no mordant; with alum, pH 7; with alum, pH 4; with alum, pH 10; with alum and iron. (Photo by author.)

FIGURE 12-7. Fresh buckthorn berries on wool. **Top to bottom:** no mordant; with alum, pH 7; with alum, pH 4; with alum, pH 10; with alum and iron. (Photo by author.)

FIGURE 12-8. The entire teasel plant—leaves, stems, and flower heads—are covered in sharp prickles. (Photo by author.)

FIGURE 12-9. Late summer teasel leaf dye on wool. **Top to bottom:** no mordant; with alum, pH 7; with alum, pH 4; with alum, pH 10; with alum and iron. (Photo by author.)

Teasel (*Dipsacus fullonum*)

Teasel is considered an invasive species in Minnesota and Wisconsin. You often find it growing along highways. The plant is a biennial. The first year it produces a rosette of leaves, usually less than a foot tall. The second year, the tall flower stalk forms. The leaves, stem, and flowers of the teasel are covered with sharp prickles, so wear leather gloves when harvesting. Both first and second year teasel leaves can be harvested to produce beige to yellow dye. Early spring leaves produce slightly greener shades (Figs. 12-8, 12-9).

Amur Maple (*Acer ginnala*)

Amur maple is a common landscape tree. It remains relatively short and produces both abundant flowers in the spring and beautiful fall foliage. It is somewhat brittle and can suffer branch breakage in storms. The leaves have the classic maple lobes (Fig. 12-10).

Fresh Amur maple bark produces a lovely shade of deep gray when simmered with iron and alum (Fig. 12-11). Bring the bark up to a gentle boil to help release the tannins. Maintain this for at least an hour. Strain the dye and add the fiber. Once the fiber is added, keep the bath at a simmer. Add

FIGURE 12-10. Amur maple leaf shows the classic maple lobes. (Photo by author.)

FIGURE 12-11. Amur maple bark on wool. **Top to bottom:** no mordant; with alum, pH 7; with alum, pH 4; with alum, pH 10; with alum and iron. (Photo by author.)

iron until the desired shade of gray is obtained. The dye needs a combination of both heat and constant stirring once the iron is added to produce an even color.

Japanese Knotweed (*Reynoutria japonica*)

Japanese knotweed is also referred to as Japanese bamboo. While the stems appear jointed, like true bamboo, it is herbaceous and dies down to the ground during the winter. The leaves are also large and somewhat heart-shaped, unlike the grassy leaves of true bamboo (Fig. 12-12). Other Latin names used for it are *Fallopia japonica* and *Polygonum cuspidatum*. Japanese knotweed is considered one of the most invasive species in the world by the World Conservation Union. It is the plant that won't die. Cutting, burning, and herbicides often kill the top, but it grows back from an extensive and far-reaching rhizome. Chopped plants should NOT be composted because they will sprout. The temperatures used in dyeing may not be hot enough to prevent sprouting. Dry the leaves after dyeing and burn them to ash.

To obtain dyes, harvest the leaves just as the flowers begin to bloom. Bring the leaves to a boil. The colors produced via different mordants and modifiers range from yellows to gold to deep green (Fig. 12-13).

FIGURE 12-12. Japanese knotweed in bloom. (Photo by author.)

FIGURE 12-13. Japanese knotweed dye on wool. **Top to bottom:** no mordant; with alum, pH 7; with alum, pH 4; with alum, pH 10; with alum and iron. (Photo by author.)

Tansy (*Tanacetum vulgare*)

Tansy is a common weed found growing throughout both North America and Europe. It is often called yellow buttons after the small, button-like flower heads (Fig. 12-14). The entire plant has a distinct, acrid smell. Tansy grows rampantly on disturbed ground.

The leaves and flower heads produce a gold/green dye that is enhanced with both copper and iron modifiers (Fig. 12-15). Simmer the chopped plant tops with water to cover. Filter the dye and add the fibers. Simmer together 30 minutes, add the iron, and continue simmering for another 30 minutes. Allow the fibers to cool in the bath. Adding a little ammonia to the cooling bath will also enhance the green color.

Norway Maple (*Acer platanoides*)

Norway maple is a large tree often used in landscaping and as a street tree due to its tolerance of urban conditions (Fig. 12-16). Its leaves and growth habit resemble the native sugar maple. Norway maple comes in a variety of cultivars. 'Crimson King' is commonly planted for its deep purple red leaves. The leaves of 'Crimson King' produce differing colors depending on season. In the early spring, just as the leaves are opening, they produce a slate blue with alum. Later season leaves give a much broader range of colors, from pink to blue-gray to brown (Fig. 12-17).

FIGURE 12-14. Tansy flower heads. (Photo by author.)

FIGURE 12-15. Tansy dye on wool. **Top to bottom:** no mordant; with alum, pH 7; with alum, pH 4; with alum, pH 10; with alum and iron. (Photo by author.)

FIGURE 12-16. Norway maple leaves produce a range of colors. (Photo by author.)

FIGURE 12-17. Norway maple 'Crimson King' dyes on wool. **Top to bottom:** no mordant; with alum, pH 7; with alum, pH 4; with alum, pH 10; with alum and iron. (Photo by author.)

To make the dye, gently simmer the fresh leaves in water. Keep the temperature low. High heat turns the dye brown. Strain the dye, then add the fiber. The dye does not need further heating. Allow the fiber to cool in the pot.

Literature Cited

Barber, E. J. W. 1991. *Prehistoric Textiles.* Princeton University Press, Princeton.

Dean, J. 1999. *Wild Color.* Watson-Guptill Publications, New York.

Mairet, E. M. 1916. *A Book on Vegetable Dyes.* Hampshire House Press, Hammersmith, London.

Vajanto, K. 2014. "Textile standards in experimental archaeology." Pp. 62–75 *in* S. Lipkin & K. Vajanto (editors), *Focus on Archaeological Textiles: Multidisciplinary Approaches. Monographs of the Archaeological Society of Finland 3.*

CHAPTER 13

Fiber Optics: Do Invasive Species Look Good on Paper?

James Ojascastro

THE DISCOVERY of paper some two millennia ago was one of the most groundbreaking innovations in human history. A versatile material, paper has been—and continues to be—used in countless ways, including in art, history, commerce, communication, construction, packaging, and hygiene. Constructed from felted plant fibers, paper was once made entirely by hand from the bark, stems, seed hairs, and leaves of a select few plant species, but since the Industrial Revolution, paper now is almost exclusively made by heavy machinery from chemically pulped wood. With concerns about the sustainability of using timber for paper, there is increasing interest in utilizing the non-wood fibers of abundant, fast-growing, and sometimes invasive, plant species.

Since both plants and paper are chiefly composed of cellulose fibers, it is not surprising that just about any plant can be rendered—whether by hand or machine—into paper. However, not all plants produce high-quality paper, and even among those that do, not all of them are worthwhile or practical to harvest for papermaking: indeed, surprisingly few species bear fibers that meet the physiological (3 mm or longer) and chemical (almost pure cellulose) criteria to manually yield strong and flexible sheets. Given these constraints, strikingly few plant species are traditionally harvested for paper: of some 400,000 plant species around the world, not more than 100 are known fiber sources in hand papermaking by tradition. Together, these few dozen traditional papermaking plants tend to share four key characteristics conducive to the production of serviceable handmade paper: they (1) have long, cellulose-rich fibers; (2) are locally abundant; (3) regenerate vigorously following harvest, either from seed or by resprouting from stumps, rhizomes, roots, and/or cuttings; and (4) are toxic. Interestingly, all these characteristics (except for the first) are common to weedy and invasive plants too. In this chapter, I will elaborate on the history of using weedy plants in papermaking traditions (Table 1), and how this 2,000-year-old precedent can help guide modern fiber artisans in responsible selection, harvest, and usage of invasive plants for paper products and paper art, including new invasives lacking historical or even contemporary precedent in papermaking (Table 2).

History
Paper Mulberry

The plant with perhaps the most outsized contribution to traditional papermaking is the paper mulberry (*Broussonetia papyrifera*), a weedy, fast-growing, and dioecious (having separate sexes) tree native to southern China. Its phloem fibers, measuring 8–12 millimeters (mm) long, are strong and flexible and yield a paper that is similarly tough but pliable (Anapanurak & Puangsin, 2001). Paper mulberry also has an impressive ability to regenerate following injury: stumps regenerate stems and leaves through their root suckers, and severed branches readily root if replaced in soil. And like most Moraceae, paper mulberry trees bear a gummy, toxic latex, which likely helps papers made from their fibers resist insect attack (Kitajima et al., 2010). Thanks to all these desirable traits for fiber-based purposes, it is not surprising that paper mulberry has a long history of human-mediated use and dispersal, and it has been transplanted chiefly by cuttings throughout Asia and Oceania to source fiber for making paper (Fig. 13-1) as well as barkcloth (beaten and felted textiles made from bark fibers). In this section, I will detail the roles paper mulberry has in different regions where it is not considered invasive to contextualize and explore its usage for paper in places where it is invasive.

In Vietnam, paper mulberry—called *dướng* in Vietnamese—is native and has been used to make paper there for many centuries (Fanchette & Stedman, 2009; Fanchette, 2016; Laroque, 2020). Exactly how long paper mulberry has been used in Vietnam for paper is not clear—in part due to

FIGURE 13.1. Paper sheets from the inner bark of seven woody species made by the author using a Western wove mould; each species represented here is considered invasive somewhere. All sheets formed using kitchen blender maceration and three hours of cooking in a sodium carbonate (Na_2CO_3) solution in ratio 1 unit Na_2CO_3 : 5 units dry bark fiber (by mass). **Left to right:** cultivated fig (*Ficus carica*); paper mulberry (*Broussonetia papyrifera*); hybrid wisteria (*Wisteria ×formosa*); mezereon (*Daphne mezereum*); burningbush (*Euonymus alatus*); white mulberry (*Morus alba*); rose of Sharon (*Hibiscus syriacus*). (Photo by author.)

ambiguity in fiber identity with the principal papermaking plant in Vietnam (dó, *Rhamnoneuron balansae*)—but it may extend back as long ago as the third century CE (Drège, 1998). In Vietnam today, paper mulberry is still used for paper, although only among ethnic groups who live in more remote mountain areas, like the Mường and Nùng (Ojascastro et al., 2024). Although techniques of cultivation, harvest, and fiber manipulation differ slightly between these ethnic groups, they generally follow the same process: first, branches are harvested; second, bark is peeled off; third, bark is scraped to remove the outer layer from the desired fibrous inner layer; fourth, inner bark is boiled in an alkaline solution; fifth, boiled bark is beaten with mallets; sixth, beaten bark is dispersed in a vat of water, usually with a mucilaginous solution that helps to make uniform sheets or paper; seventh, a porous screen is used to strain damp sheets from the vat mixture; and eighth, the sheets are dried. Once fully dry, dướng paper can be used for a variety of purposes: today, the Nùng use it mainly to package food, while the Mường use it to make artisanal products, such as notebooks and lampshades, to sell to tourists (Ojascastro et al., 2024).

In Korea, however, paper mulberry never became the predominant species used to make paper, probably because Korean artisans prized delicate paper and paper mulberry fibers are generally too coarse, and because the winters are too cold for paper mulberry to thrive under routine coppicing. So instead, the Koreans bred paper mulberry with a native but wild relative (dwarf paper mulberry, *Broussonetia monoica*) to make domesticated, cold-tolerant hybrids (dak, *B. ×kazinoki*) amenable to yearly harvest in monoculture for making hanji (Korean handmade paper). At least one of these hybrid *B. ×kazinoki* cultivars is in use in Japan, where it is called kōzō and used to make washi (Japanese handmade paper), despite possibly originating ultimately from Korea (Kuo et al., 2022).

When cultivated for papermaking, both dak and kōzō assume a bushlike, rather than treelike, growth form, and their side branches are pruned throughout the year to encourage growth of long, straight stems. Then, each November—after leaves have fallen—the stems are pruned close to the root collar at a sharp angle facing east, bundled, and steamed for about 15 minutes to facilitate easier bark peeling from the wood (Nicholas Cladis, pers. comm.). The papermaking process then proceeds much as it does in Vietnam: the peeled bark is scraped to remove the cortex, boiled in an alkaline solution prepared from wood ash, beaten with wooden mallets, and then dispersed in water with a mucilage only to be recast as wet paper sheets using a bamboo screen and wooden frame (Barrett, 1983; A. Lee, 2012). Both hanji and washi are used for painting, calligraphy, home construction (usually as shoji door screens and wall coverings), printing, weaving, and origami (Barrett, 1983).

In contrast to Japan, Korea, and Vietnam, paper mulberry in Oceania is chiefly cultivated and harvested for making barkcloth (Bell, 1985; Pang, 1992; Schattenburg-Raymond, 2020). Although barkcloth was invented independently in Africa, Asia, and the Americas, its history of usage is best studied for paper mulberry among the Austronesian diaspora, who brought cuttings of it (and a related fruit and fiber crop, breadfruit) from Taiwan and mainland Southeast Asia to Indonesia, Micronesia, Melanesia, and Polynesia (Zerega et al., 2004; Peña-Ahumada, 2020). These crops were bred over time into different cultivars, and in the case of paper mulberry, different varieties soon were tailored to suit different kinds of tapa, or barkcloth (Schattenburg-Raymond, 2020). By evaluating the genetic relationships and sex ratios of paper mulberry populations across Asia and Oceania, anthropologists have found that Austronesians dispersed varieties with desired tapa-making characteristics from island to island chiefly through vegetative propagation rather than by seed (Peñailillo et al., 2016). Some islands made tapa from a single paper mulberry variety: for example, in the South Pacific islands, genetic work has determined that all locally cultivated paper mulberries are female and share the same haplotype (Chang et al., 2015).

Other archipelagos received and cultivated multiple paper mulberry varieties, which then were tailored separately for different kinds of tapa. In Hawai'i three varieties of paper mulberry were historically cultivated and harvested for tapa (called *kapa* in Hawaiian): they are *wauke nui*, *wauke mālolo*, and *po'a'aha* (Schattenburg-Raymond, 2020). Although these three paper mulberry varieties appeared to be the principal raw material for barkcloth in Hawai'i, the bast (phloem fibers from inner bark) of several other species was also used to make tapa, including fibers from the Polynesian crop breadfruit, or *'ulu* (*Artocarpus altilis*), and from Hawaiian endemics like *ma'aloa* (*Neraudia melastomifolia*; Pang, 1992).

Sadly, with the increasing accessibility of Western clothing made by machine, tapa-making across the Pacific is now rare, and in many parts of Oceania it is now extinct, though paper mulberry persists on many islands. Fortunately, some tapa-making traditions have been revived (e.g., Hawai'i), and paper mulberry cultivars across Oceania are finding use again, for tapa as well as for paper (Schattenburg-Raymond, 2020).

Unlike in Asia and Oceania, the introduction of paper mulberry to the Western Hemisphere has been more ecologically problematic. Paper mulberry has been in the Americas since at least 1812—the year Thomas Jefferson planted several for shade at his Monticello estate (Gary, 2012). Although paper mulberry never became a predominant raw material for the American paper industry, it refused to disappear, and with both male and female trees present in North America, it instead spread prolifically outside of cultivation, especially along roadways and other disturbed areas. By 1903, paper mulberry reached Florida, and it is now ubiquitous in ruderal parts of the U.S. Southeast (Richard, 2010). With the climate warming across North America, paper mulberry continues to spread northward and is now established in Midwestern and Mid-Atlantic cities including St. Louis, Philadelphia, and New York City (Morgan et al., 2004).

Both paper mulberry and its hybrid relative kōzō are still harvested by papermakers outside of Asia (Fig. 13.1). In the United States, kōzō is cultivated in at least two locations (the Morgan Conservatory in Cleveland, Ohio; and the University of Iowa Center for the Book in Iowa City, Iowa), and both sites are carefully managed to ensure that the kōzō neither sets seed nor spreads vegetatively beyond the plots where it is grown. As for paper mulberry, American papermakers may source fiber either from cultivated individuals or by foraging from feral populations. Confusingly, American paper mulberry trees and their fibers—even those lacking any *Broussonetia monoica* ancestry—are still colloquially referred to as "kōzō," and usually accompanied by the name of the state in which they were harvested (e.g., "Alabama kōzō," "Florida kōzō," and "Texas kōzō"), and these U.S. *Broussonetia* fibers are sometimes even available commercially (Richard, 2010).

But the U.S.A. is not the only country with invasive paper mulberry that has proven useful for scalable paper arts. In Ghana, paper mulberry was introduced in 1969 through a partnership with China with the intention of building a local papermaking industry, but funding evaporated, and the planted paper mulberry began escaping into the surrounding forests (Apetorgbor & Bosu, 2011). Although the initial papermaking initiative failed, botanical and artisanal organizations started training Ghanaian artists since 2009 to exploit feral paper mulberry as a medium for local art, such as Adinkra stamp printing (Hark & Boakye, 2022). With appropriate scaling and tailoring to artisanal papers and paper products, paper mulberry especially shows the greatest promise as an underexploited, high-quality non-wood fiber resource in places like the United States and Ghana where it grows invasively, and its felling or removal as part of habitat restoration should involve partnership with local paper artists to ensure that its long, strong, and flexible fibers do not go to waste.

White Mulberry

Like paper mulberry, white mulberry (*Morus alba*) is another weedy tree from East Asia with a long history of use in traditional papermaking (Fig. 13.1). The earliest evidence of paper from white mulberry dates to the fifth century CE, in Xinjiang (X. Li et al., 2015). Although microscopic fiber analyses have some difficulty distinguishing white mulberry fibers from paper mulberry fibers, it is believed that white mulberry paper continued to be made in China and possibly Central Asia for the next millennium and a half (Cartwright et al., 2014). In 1908, archaeologist Marc Aurel Stein discovered Khotanese documents dating to the Tang Dynasty (618–907 CE), also in what is now Xinjiang; these too were made from white mulberry fibers (Cai, 2020). By the Ming Dynasty (1368–1644 CE) mulberry (likely both white and paper) fibers were being used in banknotes (Cartwright et al., 2014). Even today, papermakers in both Samarqand (Uzbekistan) and Moyu (Xinjiang, China) still make white mulberry paper by hand (X. Li et al., 2015; Rickleton, 2018).

Despite its utility in traditional papermaking, the principal historical usage of white mulberry was for rearing the silk moth (*Bombyx mori*). Silk, a more luxury fiber product than any paper, spurred investors to plant white mulberry trees in Europe and the Americas in the hopes of producing silk locally. However, most of these initiatives failed, and the white mulberry quickly escaped cultivation. Today, white mulberry is a ubiquitous, cold-tolerant weed in urban areas across eastern North America, with an impressive latitudinal breadth spanning Austin to Ottawa (plus Buenos Aires in South America; Ghersa et al., 2002). Unfortunately, the non-native distribution of white mulberry in North America largely overlaps with its native cousin, red mulberry (*Morus rubra*), whose survival it threatens due to risk of hybridization (Burgess & Husband, 2006; Petty, 2010). As white mulberry continues to spread across the United States, stands of pure red mulberry are becoming less common, and they are increasingly only found in undisturbed closed-canopy forests, where the sunlight-loving white mulberry is slower to invade (Burgess & Husband, 2006).

Although white mulberry phloem yields strong, cream-colored paper, it seems to not have been the preferred papermaking material when *Broussonetia* (either paper mulberry or kōzō/dak) was also available; this probably explains why white mulberry was grown and harvested for paper only in colder, drier climates like Xinjiang, where paper mulberry and its hybrid descendants do not grow. Additionally, physiological differences may explain why white mulberry is apparently less preferable for paper than paper mulberry: first, white mulberry fibers are about half as long (4 mm; Ojascastro, 2023) as the average paper mulberry fiber (8 mm or more), and so paper made from the latter is stronger than the former; and second, white mulberry phloem tends to produce hard, scarified inclusions even at young ages/small diameters, which must be removed to produce high-quality paper and consequently make the bark scraping process difficult and more labor intensive (Ojascastro, 2023). Better coordination between land managers, conservationists, and papermakers in regular coppicing and/or removal of white mulberry trees can facilitate achievement of mutual goals: routine harvest can both improve the quality of fiber obtained for papermakers and curb the spread of white mulberry (and by extension, endangerment of red mulberry) by decreasing the likelihood that white mulberry trees reach reproductive size.

Other Invasive Species

With fiber being a desirable material for products like textiles and paper, it is not surprising that many fiber plants are planted or have become naturalized outside of their native range; some examples include jute (*Corchorus olitorius*), hemp (*Cannabis sativa*), flax (*Linum usitatissimum*), and mitsumata (*Edgeworthia chrysantha*). But of about 50 plant species with well-documented traditions in hand papermaking, only paper mulberry and white mulberry have truly become invasive outside of their native habitat.

However, just because a plant does not have a tradition of use in hand papermaking does not make it automatically unsuitable for paper. Although most invasive plants lack traditions of use for fiber products, maverick papermakers today are demonstrating that they can still be very useful raw materials for making good-quality paper for art and even consumer products like printing paper and notebooks, albeit still at small scales. Since the invasive plants used for hand papermaking are extremely diverse, and because plant anatomy helps to inform what kinds of fibers are accessible to hand papermakers, I will organize them into four categories: phloem (which comes in two types—primary and secondary), monocot stem, leaf, and seed hairs. Note, however, that this taxonomy excludes a few organs that in rare cases can supply fibers useful for hand papermaking:

- wood (secondary xylem), which is the main raw material for commercial, machine-made paper, is too hard and brittle to process by hand for paper due to high lignin content;

- fruits (excluding any wind-dispersed seeds that may exist within) rarely provide adequately sized fiber, and when they do (e.g., coir from coconut, *Cocos nucifera*), the fiber is short, coarse, and lignified and yields poor-quality paper (Main et al., 2014); and

- roots can be difficult to extract from the ground and infrequently bear usable fiber (with rechakpa, *Stellera chamaejasme*, an herb long used in Tibetan papermaking, being the notable exception; Helman-Ważny, 2014; H. Li et al., 2014).

Phloem

Phloem, the tissue that conducts sugars produced through photosynthesis from the leaves to the rest of the plant, is the preferred source of fiber in most hand papermaking traditions around the world. This tissue comes in two varieties, depending on how it was formed: primary phloem and secondary phloem. Primary phloem, which is present in both woody and herbaceous plants, is produced by an apical meristem, the growing tip of the plant which allows stems and roots to elongate. Secondary phloem is present only in woody dicots and is created by the cambium—a thin cylinder of meristematic tissue between the xylem (wood) and phloem which allows the stem to grow in circumference. Phloem with a strongly fibrous character—regardless whether primary or secondary—is called *bast*, and this term is in common parlance among papermakers and other fiber artists. In general, when comparing bast tissues commonly used in hand papermaking, fibers deriving from primary phloem (e.g., hemp, ramie, flax) tend to be long (20 mm or more) and tightly adhered together with latex or gums, while those deriving from secondary phloem (e.g., paper mulberry, white mulberry, lokta) tend to be shorter (10 mm or less) and are readily de-gummed. As a result, primary phloem fibers often require additional steps first to remove the adhesive compounds (usually through retting—a form of fermentation) and next to chop the fibers into shorter and more uniform pieces (usually through use of a beating machine such as a Hollander beater) to avoid tangles and knots in the resulting paper sheet. In this extensive section, I will describe examples of both primary and secondary phloem fibers in invasive plants that are usable for hand papermaking.

Bark (as the fiber-rich secondary phloem tissue, from woody species) is the predominant raw material used in East Asia, Mexico, and Madagascar for traditional papermaking. In addition to white mulberry, paper mulberry, and kōzō, several dozen more species of woody plants around the world have their bark harvested by tradition for paper: these include mitsumata (*Edgeworthia chrysantha*) and gampi (*Wikstroemia sikokiana*) in both Korea and Japan (Barrett, 1983; A. Lee, 2012), wingceltis (*Pteroceltis tatarinowii*) in China (Mullock, 1995), avoha (*Gnidia* spp.) in Madagascar (Rantoandro, 1983), lokta (*Daphne bholua* and *D. papyracea*) and argeli (*Edgeworthia gardneri*) in Nepal and Bhutan (Trier, 1972; Imaeda, 1989), dó (*Rhamnoneuron balansae*) in Vietnam (Laroque, 2020), khoi (*Streblus asper*) in Thailand and Cambodia (Beckett, 1888; Boonpitaksakul et al., 2019), and over a dozen species (at least four figs, plus *Vachellia cornigera*, *Morus celtidifolia*, *Trema micranthum*, *Ulmus mexicana*, *Brosimum alicastrum*, *Sapium glandulosum*, *Urera caracasana*, and *Myriocarpa longipes*) in central Mexico (von Hagen, 1944; Peters et al., 1987; López Binnqüist et al., 2012). Despite this incredible diversity of papermaking plants all around the world, all these listed above show very similar characteristics in their bark—which in turn reflect a largely common evolutionary history.

Nearly all plants whose bark is used by tradition for papermaking fall into two categories: the urticalean rosids (families of Urticales, the order containing nettles, or *Urtica*), and the Thymelaeaceae family. Although these two categories are not closely related to each other, the woody members in both groups often display a key physiological characteristic, called "strong bark," where the bark peels off in long strips when pulled away from the wood (Gentry & Vazquez, 1993). Furthermore, by examining the bark anatomy under the microscope, woody plants with the strong bark character tend to have longer phloem fibers than woody plants lacking it (Ojascastro, 2023). With longer fibers generally correlating with paper strength (Dadswell & Watson, 1961; Tutus et al., 2010), it is not surprising to find that many species used in traditional papermaking have this "strong bark" character. By manipulating woody plants in the field, and checking for "strong bark," high-quality fiber plants can be quickly identified for modern hand papermaking (Lutz, 1983).

Thymelaeaceae Phloem Fibers

With nearly 4,800 species represented across the Thymelaeaceae and urticalean rosids, it is also not surprising to discover several that are also invasive. Within the Thymelaeaceae, or mezereon family, two have become naturalized in North America: spurge-laurel (*Daphne laureola*) in the Pacific Northwest, and mezereon (*D. mezereum*) in the northeastern U.S. Spurge-laurel, officially listed as a Class B noxious weed by Washington State, is highly toxic and reproduces prolifically by seed and by root suckers from British Columbia to Oregon (Washington State Noxious Weed Control Board, 2006). At least one papermaker from Seattle, Neal Bonham, has made sheets from spurge-laurel, and he writes that his choice to use it was inspired by its similarity to lokta (*D. bholua*), used to make paper in the Himalayas for the past thousand years (Trier, 1972). Spurge-laurel paper is smooth, soft, and white, and probably highly insect-resistant (i.e., has some toxic properties), judging from the skin and sinus irritation Bonham felt while processing the fiber (Thomas & Thomas, 1999). And on the other side of the continent, paper from another congeneric invasive—mezereon (*D. mezereum*), currently spreading in woods across upstate New York and Vermont—has been suggested (Lutz, 1983), with the prediction that such sheets would be of comparable excellence to other Thymelaeaceae-based papers like Japanese gampi. To confirm this in practice, the author made mezereon sheets in September 2022 and found the resulting paper (Fig. 13.1) of high quality: strong, thin, crisp, and translucent.

Finally, although Thymelaeaceae are mostly composed of woody species, there is a tiny herbaceous annual from Europe, annual thymelaea (*Thymelaea passerina*), which is now established in just a few populations across the American Midwest (Pohl, 1955; Vincent & Thieret, 1987; Holmes et al., 2000; Kostel, 2009). This plant is worth mentioning especially because a larger, woody congener, mitnan (*T. hirsuta*) has been used off and on since the early 1980s as a raw material by hand papermakers in Israel (Schmidt & Stavisky, 1983). Unfortunately, at present, it is doubtful that enough annual thymelaea could be harvested in the United States to make even a small amount of paper, but continued monitoring of the American populations coupled with papermaking proofs-of-concept from European populations could inform whether annual thymelaea fibers do indeed yield strong paper.

Urticalean Rosid Phloem Fibers

Nearly 4,000 plant species in four botanical families compose the urticalean rosids: the Moraceae (figs and mulberries; 1,137 species), the Ulmaceae (elms and zelkovas; 56 species), the Cannabaceae (hemp, hops, and hackberries; 117 species), and the Urticaceae (nettles and allies; 2,625 species). Although two members of the Moraceae—white mulberry and paper mulberry—were discussed at length earlier in this chapter, they are certainly not the only weedy species in this family that can be leveraged for artisan papermaking. Other invasive confamilials include (but are not limited to!) weeping fig (*Ficus benjamina*), native to Australasia but now naturalized in Florida and the West Indies; cultivated fig (*F. carica*), native to the Mediterranean but now spreading across the California Channel Islands; and Chinese banyan (*F. microcarpa*), originally from China and the western Pacific but now invasive in Florida, Central America, and Hawai'i (University of Florida, 2022). Of these, both cultivated and weeping fig have been shown conclusively to yield high-quality handmade paper. Experiments by the author have shown that the *F. carica* cultivars 'Brown Turkey' and 'Chicago Hardy' bear long (~6 mm) phloem fibers that confer great strength and crispness to the resulting pale brown sheets (Ojascastro, 2023; Fig. 13.1). Furthermore, experiments by papermaker Brad Mowers have shown that weeping fig can be used to make sheets of amate, the Mexican analog of handmade paper (Peters et al., 1987; Mowers, pers. comm.). Given these proofs-of-concept, and the fact that at least four fig species have been used to make amate historically (López Binnqüist et al., 2012), it is probable that the inner bark from any member of the genus *Ficus*—weedy and non-weedy—could be used in hand papermaking.

Elms, also members of the urticalean rosids, have strong, fibrous bark, and although the family is small, at least one is invasive and has been tested for use in hand papermaking: bark of the Siberian elm (*Ulmus pumila*), a native of Central Asia and the Russian Far East that has been introduced as an ornamental shade tree in Europe and the United States, yields a unique and extremely soft paper (Lutz, 1983).

In the Cannabaceae, the species with the most outsized role as a fiber plant is hemp (*Cannabis sativa*), an annual herb grown for its physical (and, for some cultivars, also psychoactive) utility. Hemp first saw use being spun into woven textiles, and as the textiles aged and became soiled and torn, the resulting rags were then beaten and pulped to yield paper—as was done in China some two millennia ago. A more extensive beating can transform hemp stems or hemp phloem directly into suitable papermaking pulp without the intermediate steps of spinning the fibers into textiles and wearing and tearing them into rags; this direct procedure typically uses exclusively the (primary) phloem fibers, but processing both phloem fibers and the wood-like, shorter-fibered hurd (primary xylem) is possible and can yield usable (albeit slightly weaker) paper sheets (Sam-Brew & Smith, 2017). Considerable beating (or beating and fermentation) is necessary for hemp phloem, because the fibers are coarse and very long (20 mm or more); if insufficiently beaten and chopped, hemp fibers are liable to being tangled, ultimately yielding lumpy sheets of paper (Dutt et al., 2005).

The extensive effort needed to shorten the very long and coarse primary phloem fibers in hemp, coupled with the stigma associating cannabis with drug usage, has resulted in the extinction of hemp-based papermaking traditions. For example, even though hemp was a major agricultural crop in the United States a century ago, and their descendants now grow non-natively in ditches and fields across the American Midwest, the federal government has spent millions of dollars trying to eradicate these low- or no-THC cultivars due to a lack of understanding about how different varieties are cultivated around the world for fiber versus medical uses. And continued spread of hemp elsewhere is worrisome: one assessment classified hemp as a "high invasion risk" for the U.S. state of Florida, even though it has not yet become established there (Canavan & Flory, 2019). Given drug use stigmas and ecological invasion risks, hemp paper today is made under strict regulations, usually by machine and after pulping the entire plant. Despite current constraints on its usage, the impact of *Cannabis* on the history of traditional papermaking should not be dismissed: hemp fibers have been isolated from paper money dating to the Ming Dynasty in China six centuries ago

(Cartwright et al., 2014), and hemp phloem fibers remain in use for hand papermaking today (Thomas & Thomas, 1999).

A few other invasive Cannabaceae may also be harvested to yield usable paper. Hop (*Humulus*) is an herbaceous, vining relative of hemp. The phloem of common hop (*H. lupulus*) was used to make paper in Europe during times of war (Isenberg, 1956), and it is still used occasionally to make paper by hand today (Thomas & Thomas, 1999). Although widely distributed due to its utility as a flavoring for beer, common hop does not have invasive tendencies. Its close relative Japanese hop (*H. japonicus*), however, has long escaped its initial introduction as a garden ornamental and is now widely dispersed, especially in open sun and along stream banks. It is unclear if Japanese hop has ever been used experimentally as a raw material for paper, but its ubiquity and placement in Cannabaceae suggest that paper made from its phloem would perform well.

To round out fibrous invasive Cannabaceae in this chapter, the woody genus *Trema* is worth mentioning as a candidate to explore further as a fiber source in hand papermaking. Jonote colorado (*T. micranthum*), the favored raw material for amate in central Mexico today, happens to have a close relative that is invasive in Hawaiʻi: gunpowder tree (*T. orientale*). Introduced around 1870, gunpowder tree quickly escaped extensive planting in 1925 and is now established on all five major Hawaiian islands. Its light wood was burned to make charcoal for mixing in gunpowder (hence the common name), and its bast was used for making fish nets (Little & Skolmen, 1989). In northeastern India, where *T. orientale* is native, it was used for barkcloth (Walker, 1927). Given the demonstrated use of *T. micranthum* and *T. orientale* for certain fiber products, it is worth testing specifically whether *T. orientale* bast can be co-opted for making paper as well as tapa.

The final member of the urticalean rosids, the nettle family (Urticaceae), is composed of members with a long history of use for textiles, including the herbaceous Eurasian natives ramie (*Boehmeria nivea*) and stinging nettle (*Urtica dioica*), both distributed beyond their native ranges due to their histories as fiber crops. Although ramie is believed to be one of the earliest fibers used for both textiles and paper, its coarseness and difficulty of extraction quickly led to its partial or entire replacement by more easily processable alternatives. Ramie stems are rich in sticky gums, which make isolation of the extremely long (~155 mm) fibers within extremely time- and labor-intensive, and only achievable through an extensive retting process (Maideliza et al., 2017; Hamad et al., 2019; Yao & Wei, 2021). Once isolated, ramie phloem fibers are extremely strong—stronger still when wet—and they are choice natural materials for making ropes and coarse cloths; these products in turn need to be torn, beaten, and/or retted to sufficiently shorten the fibers to be usable for paper.

Stinging nettle is an interesting case to list here, because the species appears to have a broad native distribution, inclusive of the Americas as well as Eurasia. The caveat is that there are subspecies with regional provenances: the monecious (having flowers of both sexes on one plant) subspecies *Urtica dioica* subsp. *gracilis* is native to the Western Hemisphere, while the dioecious (having separate male and female plants) subspecies *U. dioica* subsp. *dioica* is from Eurasia and is now established non-natively in the Americas. Both subspecies perform suitably for a variety of fiber products, including textiles and paper, and they face the same processing challenges as ramie: although stinging nettle fibers are on average a little shorter (but ranging between 50 and 210 mm in length), they are similarly strong and tear resistant and therefore still require retting before weaving or papermaking (Bacci et al., 2009; Beisinghoff, 2010; Rivers, 2020). Despite being labor intensive, using stinging nettle for paper and other fiber products can be ecologically worth the effort: nettles require far less water and pesticides than cotton in cultivation (Beisinghoff, 2010).

Flax Phloem Fibers

Like hemp, common flax (*Linum usitatissimum*) is an annual that has been in cultivation for millennia as a source of fiber for paper, first via rags, and it is still a dominant raw material for hand papermaking in the Western tradition (Barrett, 2021). In a similar vein, it would be worth exploring whether any of the other flax species beyond *L. usitatissimum* do in fact yield strong paper, but few of these have been dispersed outside of their native ranges and none appear to have strongly weedy or invasive tendencies. Both European blue flax (*L. perenne*) and its American congener Lewis flax (*L. lewisii*) are very hardy herbs that sometimes appear as weeds in lawns and home gardens across the United States, and both flaxes would be worthy candidates for experimental paper should sufficient amounts of stem tissue be collected.

Legume Phloem Fibers

Although plant systematics provides a useful guide for choosing plants with long, flexible fibers in their phloems, it does not automatically mean plants outside of the Thymelaeaceae, urticalean rosids, and Linaceae cannot have their phloems harvested and beaten to make paper by hand: additional ethnobotanical context can also be used to inform the feasibility of novel or poorly attested plant uses. In terms of paper, suitability of a given plant can be supported by evidence of use for other fiber traditions, like weaving, cordage, and basketry. This is apparent among the Fabaceae (legumes), which together compose a very large and diverse family of about 20,000 species, very few of which are considered fiber plants. Among the very few whose bark is used for fiber products, two perennial vining legume genera stand out: kudzu (*Pueraria*) and wisteria (*Wisteria*). In their native East Asia, the inner bark of both genera has historically been used for weaving cloth (Bell, 1995; Mitich, 2000). Kudzu bark also has long traditions of use in basketry, and it is still used for this purpose in Japan and now also the United States (Mitich, 2000). Furthermore, several sources claim historical usage of wisteria bark for papermaking, although no material evidence of this survives (Bell, 1995; Yum, 2008). But kudzu and wisteria also share a more insidious claim to fame: both have become severely invasive outside of their native range.

In the United States, the story and identity of kudzu is complicated. Although the genus *Pueraria* has some 15 or 20 species, at least three (*P. montana*, *P. edulis*, and *P. phaseoloides*) have been introduced to the United States, and many American populations are composed of some hybrid mixture of these three (Jewett et al., 2003). American kudzu first appeared in a Philadelphia exhibition from Japan in 1876 and was advertised as an ornamental. In the early 20th century, kudzu was also marketed as a proteinaceous cattle feed, and by the Dust Bowl era, it was sold for erosion control. By the 1950s, the United States Department of Agriculture (USDA) stopped recommending kudzu as ground cover, but the damage had been done, and the vine continued to spread. By 1997, it was placed on the Federal Noxious Weed List, and today kudzu blankets some three million hectares of land, mostly in the southeastern United States. Its spread has lasting economic costs: utility companies spend $1.5 million per year to mitigate damage to power lines caused by kudzu, and land managers spend an average of $5,000 per hectare for kudzu removal (Forseth & Innis, 2004).

Given this massive ecological impact, some artisans and entrepreneurs are looking to recoup the costs of kudzu control by promoting the many ethnobotanical uses of kudzu, including harvesting the roots for starch and herbal tinctures, flowers for jelly, foliage for animal feed, and bark for baskets and (both handmade and machine-made) paper. Concerning the suitability of kudzu for paper, experiments by the Korean paper industry show that pulped whole stems and roots yield shorter fibers (~1 mm) and weaker resulting paper for kudzu as compared to softwood pulp (Kim et al., 2010;

J. Y. Lee et al., 2018). On the other hand, investigations by Lilian Bell determined that stronger paper can be made if only the phloem fibers (2–3 mm in length) are used, although it is not clear whether her remarks and measurements apply to herbaceous first-year growth (i.e., kudzu primary phloem fibers) or to older, semiwoody stems (i.e., secondary phloem; Bell, 1995). Consequently, optimizing an economical management of kudzu therefore likely depends on leveraging a combination of uses—culinary, medicinal, and artisanal—that may vary depending on the age and anatomy of the vines.

Like kudzu, the invasion of wisteria in the United States involves hybridization. Both Chinese (*Wisteria sinensis*) and Japanese (*W. floribunda*) wisteria were introduced as ornamentals, and both widely planted across eastern North America for their fragrant, trailing, purple inflorescences. However, both species can reproduce sexually with each other, yielding hybrid offspring (*W. ×formosa*) of intermediate physiology that have a remarkable ability to climb and smother trees in disturbed forests. A recent genetic study determined that over 80% of feral wisteria sampled from Charleston, South Carolina, and Tallahassee, Florida, were of hybrid origin (Trusty et al., 2007). However, given its toxicity and comparatively smaller ecological impact, invasive wisteria has fewer ethnobotanical uses explored than for invasive kudzu. Modern hand papermakers have identified fibers of intermediate length (1.3–3.7 mm long) in wisteria bast, and they in turn can yield a white paper with average tearing resistance and folding strength (Lutz, 1983; Bell, 1995), although experiments by the author yielded a textured paper with a chocolate-brown color, perhaps due to the usage of older inner bark (Ojascastro, 2023; Fig. 13.1). But what is remarkable is the absorbance: when sumi (Japanese calligraphy ink) is brushed on wisteria paper, the ink pools in attractive feathery patterns along the stroke margins (Lutz, 1983). These experiments show that wisteria control in eastern North America could theoretically support the small-scale production of artisanal painting and calligraphy papers here in the United States.

There remains one other weedy legume to mention here—it is not woody, but it is nevertheless a bona fide raw material for a centuries-old papermaking tradition: sunn hemp (*Crotalaria juncea*). Sunn hemp, an annual forb native to the Indian Subcontinent and Indochina, is the raw material of choice for Mohammed Hussein Kagzi, an artisan living in Rajasthan who is one of the last living hand papermakers following the Islamic tradition. In addition to paper, sunn hemp's long (7 mm) phloem fibers can also be used to make ropes and nets (White & Haun, 1965; Brink & Achigan-Dako, 2012; Chaudhary et al., 2015). And beyond fiber applications, sunn hemp has diverse agricultural uses: it produces allelopathic chemicals that can inhibit growth of agricultural weeds, and it also bears nitrogen-rich foliage that, when composted into green manure, can enrich soils for growing other crops. This former agricultural utility—as a natural weed inhibitor—ironically also enhances its invasiveness, and sunn hemp can now be found on farmland across six continents. In cooler agricultural regions, winter limits the invasiveness of sunn hemp because it dies from frost before being able to set seed, but in warmer places it can readily proliferate, and governments like the state of Arkansas have listed sunn hemp as a noxious weed (Sheahan, 2012). Presently, excess or escaped sunn hemp that cannot be composted could be used as a pulpwood alternative in papermaking, whether by hand or by machine, but only Mohammed Hussein Kagzi seems to be exploiting it artisanally for paper today.

Milkweed Phloem Fibers

Several other taxa not closely related to other forbs used in traditional papermaking show promise as sources of primary phloem fiber for paper. Perhaps the best examples of this are the non-succulent Asclepiadoideae, many of which have long, strong, flexible, highly cellulosic fibers in their phloem that are already used to make carpets, ropes, thread, and nets (Chauhan et al., 2013; Maji et al., 2013). Several milkweed species in the genera *Asclepias* and *Calotropis* have become established weeds outside of their native ranges, such as the eastern North American common milkweed (*A. syriaca*) in Europe, and the West Asian and North African apple-of-Sodom (*C. procera*) in the Neotropics and Australia. Individual hand papermakers have already shown that phloem fibers from species in these two genera can yield strong sheets, and although their latex content can render processing challenging (Bell, 1995; Hiebert, 2006; A. Lee, 2022), milkweeds can even be sufficiently abundant to harness for small-scale hand papermaking industries (Chauhan et al., 2013).

Malvaceae Phloem Fibers

One more family has an extensive tradition of use for fiber products: the mallows, or Malvaceae. This large family, which includes chocolate, okra, cotton, and hibiscus, is also characterized by having medium to long phloem fibers. Although cotton is by far the most globally important malvaceous fiber crop, and its phloem fibers are usable experimentally for paper (Ververis et al., 2004), its most usable fiber comes from its seed hairs, and will be described further later in this chapter. Instead, the most important malvaceous phloem fiber is jute (*Corchorus*), of which two species are important agricultural crops (*C. capsularis* and *C. olitorius*), both native to Southeast Asia and the Indian Subcontinent (and the latter with a native range extending to sub-Saharan Africa). Commercial jute is mostly used for twine and coarse textiles like burlap, but it has also been shown to be usable for handmade paper, both historically and contemporarily (Isenberg, 1956; Dutt et al., 2005; T. Li, 2018). A third mallow, kenaf (*Hibiscus cannabinus*), native to sub-Saharan Africa and now cultivated across Indomalaya and in the U.S., has seen similar historical uses for twine and cloth and is now grown industrially to make polymer composite materials and as a non-wood raw material for paper (Isenberg, 1956).

All three of these malvaceous fiber crops have occasionally escaped cultivation outside of their native ranges, but none are considered invasive. Fortunately, their utility provides precedent for testing the suitability of other Malvaceae for fiber, including species that are considered invasive, such as wireweed (*Sida acuta*), Cuban jute (*S. rhombifolia*), heartleaf sida (*S. cordifolia*), velvetleaf (*Abutilon theophrasti*), and caesarweed (*Urena lobata*), all of which are now globally distributed warm-temperate or tropical weeds (Isenberg, 1956). Fiber chemistry and physiology measurements have shown that *Sida*, *Abutilon*, and *Urena* have flexible, cellulose-rich fibers comparable in physiology to kenaf and jute (Rowell et al., 2000; Olotuah, 2006; Reddy & Yang, 2008; Sharma et al., 2013), and experiments by papermakers have shown that they all yield usable paper (Bell, 1995). One last species of Malvaceae is worth mentioning. Rose of Sharon (*Hibiscus syriacus*), a West Asian ornamental shrub planted throughout temperate regions and occasionally found as an escaped weed (e.g., in the eastern United States), has been shown repeatedly to yield usable—albeit weak, in this author's experience—handmade paper (Georgia Exotic Pest Plant Council, 2006; Hiebert, 2006; Ojascastro, 2023; Fig. 13.1).

Other Invasive Phloem Fibers

Finally, even plants lacking ethnobotanical or evolutionary precedent may be used in traditional papermaking. Even when restricting to woody species with appreciable bark tissue, this category still includes a vast number of plant species. Although very few bark fiber measurements have been made across the plant kingdom, it is probable that most of these either have no phloem fibers, or if so, the fibers are rather short (2 mm or shorter) and would therefore yield weak or brittle paper. Nevertheless, choosing to employ short-fibered invasive species for papermaking can still be purposeful and justifiable. One reason might be sheer abundance: harvesting a ubiquitous invasive could yield a considerable amount of fiber, whose properties as paper can be enhanced through the addition of some amount of a different longer or otherwise higher-quality fiber. Another reason is that the invasive plant may have unique qualities that could result in interesting (though probably not strong) sheets. And thirdly, using invasive species can make a statement, by transforming an ecologically damaging agent into a useful product or an inspirational artwork. Although there are now many examples of invasive species whose phloem fibers have been processed into handmade paper (Table 1), I will describe in detail just three species in this category that exemplify these considerations: castor bean (*Ricinus communis*), burningbush (*Euonymus alatus*), and tree of heaven (*Ailanthus altissima*).

Castor bean, native to the Horn of Africa, is now widely distributed in tropical and even warm temperate regions throughout the world. Depending on climate, castor bean may grow as an herbaceous annual (temperate regions) or as a tree (tropical regions), and thriving in ruderal, human-altered habitats in both. Consequently, castor bean is considered invasive in many places, including California, Florida, the Caribbean, and Australia (Gordon et al., 2011; Puran & Ronell, 2014). It is widely distributed in part due to its utility as a source of castor oil, which is extracted from the seeds through pressing and used historically and now in traditional medicine, food preservation, and the manufacture of perfumes, waxes, and soaps. Experiments by Luis Torres and Alberto Valenzuela at their studio, Papel Oaxaca, in San Agustín Etla, Mexico—where castor bean is invasive and locally abundant—have demonstrated that the primary phloem of young castor bean stems peels readily from the primary xylem and can be scraped, cooked, and beaten to make handmade paper (Torres & Valenzuela, pers. comm.). However, fibers in the stem have been shown to be quite short (~0.8 mm; Saeed et al., 2018), so further research is needed to demonstrate the strength and durability (or lack thereof) of castor bean phloem fiber paper.

Burningbush, a shrub native to East Asia, has been planted extensively in the United States as an ornamental hedge plant, predominantly for its attractive foliage which turns bright crimson in autumn; these leaves are readily visible in forest understories across the northeastern United States. Under typical pruning conditions, burningbush grows short, stout branches supporting many smaller twigs bearing raised, rectangular corky ridges. Larger branches bear a thick and very white phloem, which separates from the wood easily but rarely separates as one piece. No historical fiber uses are known for burningbush, and the phloem fibers come up short, averaging about 0.6 mm in length (Ojascastro, 2023). To minimize shortening the fibers further, recipes for processing burningbush phloem are therefore relatively simple: one excludes alkaline cooking entirely and involves hand-beating the raw fiber for just 15 minutes (Lutz, 1983). By this method, burningbush bast yields a crisp and very white paper with low strength and low absorbency (Lutz, 1983). However, if burningbush phloem fiber is cooked and beaten more thoroughly, the paper becomes very weak and turns a pinkish tan color, albeit with some appealing inclusions of long, unbeaten fiber bundles (Ojascastro, 2023; Fig. 13.1). The weakness and poor absorbency of burningbush fiber probably limit its usage as paper to decorative non-painting and non-writing uses like embossing or scrapbooking, unless mixed with longer and stronger fibers.

Tree of heaven, also a native of East Asia, was first introduced to Europe and North America in the 18th century. Fast-growing and readily root-suckering, it can capitalize on disturbed areas with open sun while secreting compounds that discourage the growth of other competing plants nearby. Today, tree of heaven is present on every continent except Antarctica, and it is nearly impossible to eradicate. And although tree of heaven has many uses in its native China—including for cabinetry, silk production (as a host plant for the silk moth *Samia cynthia*), and Traditional Chinese Medicine (TCM)—these uses often fail to be practical or economical elsewhere due to the lower demand for TCM outside of China and the availability of higher-quality alternative species for cabinetmaking and sericulture. The same challenge applies to paper: both wood and inner bark have been tested for use in both industrial and manual papermaking, but neither method seems to yield strong paper (Ferreira et al., 2013) due to the shortness of fibers in both wood (0.7–0.8 mm) and phloem (1.4 mm) (Baptista et al., 2014; Campagna et al., 2017). Hand papermakers who have used tree of heaven bark for paper report additional challenges and impracticalities in processing, from the thinness of the phloem layer to the bad odor that results from cooking it (Petty, 2010; Cappiello, 2021). Furthermore, although printing is possible on 100% tree of heaven bark paper, results are enhanced if the tree of heaven bark fibers are mixed with longer fibers such as cotton (Cappiello, 2021). Despite these difficulties, repurposing invasive but perhaps impractical papermaking species like tree of heaven in contemporary paper arts exhibitions is a powerful symbolic reclamation that encourages and inspires people to become more engaged in nature and ultimately better stewards of their environment (Singleton, 2012).

Monocot Stem Fibers

Since xylem and phloem tissue are scattered throughout the stems of monocots, they are not readily separable by hand papermakers; consequently, for caulescent monocots like grasses and sedges, entire stems tend to be crushed and cooked when transformed into handmade paper. Although whole stems of many herbaceous dicots may be harvested, cooked, and beaten to yield paper (Table 1), here I will focus principally on stem papers from invasive graminoid monocots, which generally tend to perform much better in papermaking case studies. Note that this taxonomy also excludes a small group of invasive non-graminoid monocots like white ginger (*Hedychium coronarium*), which have documented uses in experimental hand papermaking (Table 1; Torres & Valenzuela, pers. comm.).

Graminoid stems may have been among the first fibers to be incorporated into paper, chiefly as agricultural byproducts: sugarcane (*Saccharum officinarum*), wheat (*Triticum aestivum*), and rice (*Oryza sativa*) are grown primarily for culinary use, and after the desired grains (or in the case of sugarcane, juices) are extracted, the remaining fibrous waste can be recycled for secondary purposes, including paper. Although fibers from both wheat and rice have been identified in Ming Dynasty-era Chinese paper banknotes (Cartwright et al., 2014), this historical precedent as fibers has allowed modern hand papermakers to co-opt other cereal grains for paper too, including corn (*Zea mays*), barley (*Hordeum vulgare*), and sorghum (*Sorghum bicolor*) (Bell, 1995; Thomas & Thomas, 1999; Kiaei et al., 2011; Saeed et al., 2017; Torres & Valenzuela, pers. comm.). Non-cereal grasses may also be used to make paper, and there is historical precedent: esparto (*Macrochloa tenacissima*) was used to make paper in Spain, and sabai (*Eulaliopsis binata*) is still mixed with lokta (*Daphne bholua*) fiber to make paper in Nepal, but none of these are considered invasive anywhere. But with large, bunch-forming grasses becoming problematic invasives in many regions, there is now increasing interest in using fibers from invasive graminoids in lieu of wood fibers for both handmade and machine-made paper.

Good examples of invasive grasses with demonstrable use in papermaking include giant reed (*Arundo donax*), common reed (*Phragmites australis*), and pampas grass (*Cortaderia selloana*) (Bell, 1995; Thomas & Thomas, 1999; Marques et al., 2010). Sedges—related graminoids in the Cyperaceae family—can also bear usable stem fibers: purple nutsedge (*Cyperus rotundus*), native to the Eastern Hemisphere and invasive in the Western Hemisphere, has been tested as a raw material for making paper, paperboard, and cardboard. However, the resulting sedge paper is 10 times weaker than commercial papers like rag bond and newsprint on account of short fibers (0.7 mm) in the *C. rotundus* stem (Bidin et al., 2015). Indeed, among graminoids broadly, stem fibers are relatively short (usually 3 mm or less in length) and are rich in silica—both qualities that result in weaker, lower-quality paper (Ilvessalo-Pfäffli, 1995). Consequently, paper from invasive monocot stems may best be suited for products like newsprint, disposable paper packaging, and crafts where high-quality paper is not needed.

Leaf Fibers

Although nearly all plants bear leaves—organs which evolved to conduct photosynthesis—surprisingly few species bear leaves that yield usable paper. The principal (although not exclusive) examples of these are monocots, especially abaca (*Musa textilis*), a banana native to the Philippines that is now cultivated pantropically as a fiber crop. Bananas in particular bear exemplary leaf fibers because of their aboveground architecture: since bananas lack true stems, the petioles instead are reinforced with long fibers to support the total aboveground weight of the herb. These petioles are the source of the famous "Manila hemp" fibers that have been extracted and used for centuries to make garments, ropes, and paper. Congeners are also exploited in papermaking today (Miller, 2010), and artisans like Fabian Correa Gómez harvest, cook, and beat fibers from cultivated bananas to make *agua papel*, a crisp paper prized as a top-quality medium for folding origami (Terry, 2022).

Although abaca itself is not invasive even in cultivation, several leafy monocots with promising or demonstrated applications for papermaking are. Perhaps the most notorious invasive of these is the water hyacinth (*Pontederia crassipes*), a globally invasive wetland plant native to South America. Water hyacinth leaf fibers, which measure 1.6 mm in length on average (Goswami & Saikia, 1994), are long enough to be used for fiber products, and they have been used industrially as a non-wood raw material for making paper and board in both India (Ghosh et al., 1984) and Indonesia (Joedodibroto et al., 1983). Artisanally, water hyacinth is used for weaving baskets and floor mats in Central Cambodia, and at least one papermaker in Vietnam makes water hyacinth paper by hand (Lý, pers. comm.). Water hyacinth is also a major threat to wetlands in the United States, especially along the Gulf Coast, and American artists like Megan Singleton have harvested and cooked water hyacinth to yield a dark pulp for painting large, meandering compositions reminiscent of Louisiana bayous—though she notes that water hyacinth–based compositions of this size need structural reinforcement with longer, stronger fibers such as abaca (Singleton, 2012). With water hyacinth being so abundant around the world, workshops and apprenticeships can be expanded to allow artisans-in-training to earn an income exploiting and creating with this otherwise ecologically deleterious wetland plant.

Another group of fibrous-leaved plants with a history of use in fiber products are the agaves (*Agave* spp.), of which sisal (*A. sisalana*) is by far the most widely cultivated. Native to the Yucatan Peninsula, sisal was exported widely during the 19th century, and sisal plantations popped up in East Africa, Southeast Asia, Oceania, and Florida. While some plantations still grow, process, and export sisal, other regions shifted their economies, and descendants of agricultural sisal now grow feral in places like Florida, where the species is now listed as a Category II invasive (Brown, 2002). Today, sisal finds occasional artisanal use, and it was a raw material of choice when hand papermaking was first introduced to South Africa during the 1980s (Marshall, 2006). Fiber comparisons between sisal and other congeners reveal comparable microphysiologies, with lengths ranging generally between 1 and 3.5 mm (McLaughlin & Schuck, 1991) and yielding papers as strong or stronger than kraft wood-pulp paper (McLaughlin, 2000). Furthermore, these fiber morphology studies imply that century plant (*A. americana*), a North American native now established or invasive on the remaining five habitable continents, must be an underexploited raw material for paper and rope: its leaf fibers measure 1.7 mm on average—only slightly shorter than sisal (1.84 mm).

These conclusions seem to hold broadly for xeric (dry-habitat) Asparagaceae as a whole: leaf fibers from the genus *Yucca* measure 1.8–3.5 mm in length, and papermaking projects with non-invasive yuccas like *Y. baccata* have produced tawny-colored, textured sheets (McLaughlin & Schuck, 1991; Thomas & Thomas, 1999). However, more refined papers are possible if lye (sodium hydroxide, NaOH) is used to cook the fibers: with a strong enough caustic, papermaker Lillian Bell has shown that fibers from common yucca (*Y. filamentosa*), a North American native now established in parts of Europe (Szatmari, 2012), yield good-quality cream-colored papers regardless of what sheet formation technique is used (Bell, 1995; Hiebert, 2006). Leaf fibers from the less closely related confamilials *Nolina* and *Dracaena* also seem to have suitably sized cellulosic fibers for papermaking (McLaughlin & Schuck, 1991); snake plant (*D. trifasciata*), a West African native now encountered globally as a houseplant, garden plant, and occasional weed (Krauss, 2012), was historically used for bowstrings but also performs suitably for artisanal paper (Bell, 1995).

A few other non-Asparagaceae invasives yield usable fibers for hand papermaking. For example, Eurasian ornamentals like orange daylily or ditch lily (*Hemerocallis fulva*) and yellow flag (*Iris pseudacorus*) have escaped cultivation and become invasive in the eastern United States, and both bear fibers that can be scraped, cooked, beaten, and cast into paper by hand (Thomas & Thomas, 1999); in the case of the latter species, the resulting paper is even absorbent enough for printmaking (Cappiello, 2021). Other cormaceous monocots with large leaves—like taro (*Colocasia esculenta*), a starch crop native to South and Southeast Asia now invasive in Spain and parts of the U.S.—can be used for paper, although resulting sheets from taro fibers specifically are reportedly stiff, and the pulp can be a skin irritant due to presence of oxalic acid raphides (Dana et al., 2017; Mowers, pers. comm.). And although monocots seem to yield the best leaf fibers, dicot leaf tissue should not be discounted: artist Mary Tasillo has harvested princesstree (*Paulownia tomentosa*) petioles to make papers for letterpress printing (Tasillo, 2010). With leaves—especially dicot leaves—being comparatively underexplored as a raw material for papermaking, further experiments are necessary to demonstrate their artistic and even utilitarian suitability for paper and paper products.

Seed Fibers

Some plants bear seeds that have fluffy tufts of fibers attached, which serve to catch wind and carry the seed wherever the air currents take it. Since these fibers are long and rich in cellulose, this kind of wind-dispersal architecture can be co-opted by papermakers to yield strong paper sheets. Essentially all seed-hair fiber plants used in hand papermaking are in one of just three families: Malvaceae, Apocynaceae, and Asteraceae.

As mentioned earlier, the dominant seed-hair fiber plant in both the textile industry and in hand papermaking is cotton (*Gossypium* sp.), a malvaceous shrub that bears long (10–40 mm) fibers attached to and enveloping its seeds (Rowell et al., 2000). Its confamilial kapok (*Ceiba pentandra*) is a tall, spike-covered tree that disperses similarly and bears cellulose-rich fibers of comparable length (8–30 mm) to cotton. Both cotton and kapok, although widely cultivated beyond their native ranges, seem to show little invasive tendency anywhere (Mendes-Rodrigues et al., 2019; Hending et al., 2021).

Better examples of invasive plants bearing paper-worthy seed hairs are present in the Apocynaceae, and these have also been mentioned earlier as sources of phloem fiber: milkweeds in the genus *Asclepias* and *Calotropis*. Handmade papers from both kapok and milkweed seed hairs are strong and especially soft and white, and such sheets are easily made without needing much (if any) beating or cooking, because the fibers are already cellulose-rich and largely separated excepting the point where they connect to the seed (Rich, 2022; Torres & Valenzuela, pers. comm.). The drawback for papermakers, however, is that separating the seeds from the fluff can be quite time-consuming (Petty, 2010).

The family not yet mentioned much in this chapter is the daisy family (Asteraceae), a very large clade represented by almost 30,000 species. Some members of the Asteraceae, such as the thistles (*Cirsium* and *Carduus*), reproduce by airborne seeds held aloft in the wind by cellulosic hairs. These hairs tend to be much shorter than those of kapok, cotton, or milkweed, but they still can be harnessed for artisanal papermaking. Experiments by papermaker Maureen Richardson have shown that bull thistle (*Cirsium vulgare*) seed hairs will yield a brown, textured paper that is attractive but not particularly strong (Hiebert, 2006). Sheets made by Gin Petty from nodding or musk thistle (*Carduus nutans*) are similarly textured—especially if the fiber is processed without removing seeds (Petty, 2010). These strength and texture considerations need to be borne in mind when deciding how such thistledown papers are to be used (Petty, 2010).

Despite being short-fibered and yielding delicate paper, many Asteraceae species (including both *Cirsium vulgare* and *Carduus nutans*) are both locally abundant and widely distributed weeds—an advantage to papermakers given the large amount of fiber needed even for a small stack of paper. Like dicot leaf fibers, Asteraceae seed fibers are therefore an underexploited raw material for papermaking, and more papermaking experiments should be conducted to utilize fibers from more weeds, including (but not limited to) tall goldenrod (*Solidago altissima*), which is invasive in Europe (Dudek et al., 2016), and common dandelion (*Taraxacum officinale*), which is weedy almost everywhere.

Conclusion

As humans continue to move themselves and other organisms around the planet, invasive species will continue to be a major concern for the long-term integrity of ecosystems. Biological control of invasive species is a globally expensive endeavor that is expected to only increase in cost. Furthermore, coppicing and culling weedy species often involves challenges about how to treat or dispose of the waste biomass without perpetuating or exacerbating their spread, or that of associated insects, fungi, or other pathogens, via seeds, spores, eggs, and/or cuttings. Recycling biomass from invasives for utilitarian products and art can help alleviate pressure regarding what to do with ecologically detrimental species, but such co-optation levies important responsibilities on the harvester, which fall in two main categories (Johnson, 2010):

- **Educate.** Before any processing or removal is to be done, it is your responsibility to familiarize yourself with the species and landowners in your area. If you don't recognize the plants you hope to use for art, reach out to local botanists and naturalists for help. Talk to neighbors and landowners to learn about what plants grow in your area, and what invasive plants might threaten your local ecosystem. Familiarize yourself with any botanical hazards, as many invasive plants may bear irritating compounds (e.g., mezereum), physical defense (e.g., multiflora rose, *Rosa multiflora*), or both (e.g., stinging nettle). Study how invasive plants defy efforts to remove or control them—many can readily regenerate from fragments still left in the ground. And as you continue to learn, pass along your knowledge to educate others to let them know where invasive species are and how to properly control or remove them.

- **Contain.** Invasive species take advantage of ecosystems through dispersal—sometimes accidental, sometimes deliberate—by people. Once established, invasive species become frustratingly difficult to eradicate, and short of total removal, most management strategies focus on containing invasives and limiting their further spread. While invasive species often present as an abundant and underexploited resource, artists who use tissue from invasive species also have a moral responsibility to do no further ecological harm in their harvesting activities. Harvest plants before they set seed (as seeds are the primary dispersal mechanism for most plants) and strive to collect entire plants by uprooting them entirely where possible. Contain collected tissue securely by using thick-walled garbage bags, and never transport harvested tissue lest the harvested species become established and invasive elsewhere. To discard any unusable tissue, enclose it securely in thick-walled bags and toss in the garbage.

While creating art from ecologically damaging species is a rewarding experience, it is impossible to expect restoration ecology and land management to proceed effectively purely by the artist's hand. Invasive species are pernicious and ubiquitous, and managing their spread requires cooperation by many people—ecologists as well as laypersons—working toward a common goal from different experiences and vocations. Creating art from invasive plants is not a silver bullet solution to a complex and difficult problem, but rather a creative complement to the systematic and scientific efforts of restoration ecologists and land managers. Moreover, the effectiveness of art in depicting the challenges caused by invasive species is not limited to just the usage of invasives as a raw material. Art is a performance—requiring the engagement and interaction of an audience in response to the actions of the artist. In creating paper and paper art from invasive species, the principal message is not a material one—that the invasive plants can be cut back or removed to make art. The main takeaway is that art can be a powerful megaphone to encourage the public to be aware and engaged stewards of the places they live—not just through invasive species control, but also in other sustainable ways, including growing native plants, eating local foods, sourcing energy renewably, taking public transit, and the like.

Literature Cited

Anapanurak, W. & A. Puangsin. 2001. "Paper mulberry pulp properties by various alkaline processes." Pp. 309–316 *in* Proceedings of the International Symposium on Paper Mulberry and Hand-Made Paper for Rural Development. Kasetsart University, Bangkok.

Apetorgbor, M. M. & P. P. Bosu. 2011. "Occurrence and control of paper mulberry (*Broussonetia papyrifera*) in southern Ghana." *Ghana Journal of Forestry* 27: 40–51.

Babcock, M. 2017. "Papermaking in Detroit: The Invasive Paper Project by Megan Heeres." *Paper Slurry*, 7 March 2017. https://paperslurry.com/blog/2017/03/07/papermaking-in-detroit-the-invasive-paper-project-by-megan-heeres

Bacci, L., S. Baronti, S. Predieri & N. di Virgilio. 2009. "Fiber yield and quality of fiber nettle (*Urtica dioica* L.) cultivated in Italy." *Industrial Crops and Products* 29: 480–484.

Baptista, P., A. P. Costa, R. Simões & M. E. Amaral. 2014. "*Ailanthus altissima*: An alternative fiber source for papermaking." *Industrial Crops and Products* 52: 32–37.

Barrett, T. 1983. *Japanese Papermaking: Traditions, Tools, and Techniques*. Weatherhill, New York and Tokyo.

Barrett, T. 2021. *European Hand Papermaking: Traditions, Tools, and Techniques*. Legacy Press, Ann Arbor.

Beckett, W. R. D. 1888. "Streblus paper." *Bulletin of Miscellaneous Information* (Royal Botanic Gardens, Kew) 1888: 81–84.

Beisinghoff, B. 2010. "Tales of nettles." *Hand Papermaking* 25(1): 37–38.

Bell, L. 1985. *Papyrus, Tapa, Amate, & Rice Paper: Papermaking in Africa, the Pacific, Latin America, and Southeast Asia.* Liliaceae Press, McMinnville, Oregon.

Bell, L. 1995. *Plant Fibers for Papermaking.* Liliaceae Press, McMinnville, Oregon.

Bidin, N., M. H. Zakaria, J. S. Bujang & N. A. Abdul Aziz. 2015. "Suitability of aquatic plant fibers for handmade papermaking." *International Journal of Polymer Science* 2015: 165868.

Boonpitaksakul, W., K. Chitbanyong, B. Puangsin, S. Pisutpiched & S. Khantayanuwong. 2019. "Natural fibers derived from coi (*Streblus asper* Lour.) and their behavior in pulping and as paper." *BioResources* 14: 6411–6420.

Brink, M. & E. G. Achigan-Dako. 2012. *Plant Resources of Tropical Africa.* Wageningen University, Wageningen.

Brown, K. 2002. "*Agave sisalana* Perrine." *Wildland Weeds* 2002(Summer): 18–21. https://www.se-eppc.org/wildlandweeds/pdf/summer2002-brown-pp18-21.pdf

Burgess, K. S. & B. C. Husband. 2006. "Habitat differentiation and the ecological costs of hybridization: The effects of introduced mulberry (*Morus alba*) on a native congener (*M. rubra*)." *Journal of Ecology* 94: 1061–1069.

Cai, M. 2020. "Overview of paper and papermaking in Xinjiang, China." *Z Badań nad Książką i Księgozbiorami Historycznymi* 14: 411–425.

Campagna, M. N., M. Gattuso, M. L. Martinez, M. V. Rodriguez & O. Di Sapio. 2017. "Novel micromorphological features of wood and bark of Argentinean Simaroubaceae." *New Zealand Journal of Botany* 55: 134–150.

Canavan, S. & S. L. Flory. 2019. "UF/IFAS Industrial Hemp Pilot Project: Invasion risk." University of Florida. https://programs.ifas.ufl.edu/media/programsifasufledu/hemp/invasion-risk-hemp-fact-sheet.pdf

Cappiello, J. 2021. "Art professor gets creative with invasive species." University of Louisville. https://louisville.edu/artsandsciences/news/all/rachel-singe-invasive-species; https://www.uoflnews.com/section/arts-and-humanities/uofl-art-professor-gets-creative-with-invasive-species/

Cartwright, C. R., C. M. Duffy & H. Wang. 2014. "Microscopical examination of fibres used in Ming dynasty paper money." *The British Museum Technical Research Bulletin* 8: 105–116.

Castañuela, Z. 2010. "Papermaking and fresh eggs: Three paper samples." *Hand Papermaking* 25(2): 27.

Chang, C.-S., H.-L. Liu, X. Moncada, A. Seelenfreund, D. Seelenfreund & K.-F. Chung. 2015. "A holistic picture of Austronesian migrations revealed by phylogeography of Pacific paper mulberry." *Proceedings of the National Academy of Sciences* 112: 13537–13542.

Chaudhary, B., M. K. Tripathi, H. R. Bhandari, S. K. Pandey, D. R. Meena & S. P. Prajapati. 2015. "Evaluation of sunn hemp (*Crotalaria juncea* L.) genotypes for high fibre yield." *Indian Journal of Agricultural Sciences* 85: 850–853.

Chauhan, S., A. K. Sharma, R. K. Jain & R. K. Jain. 2013. "Enzymatic retting: A revolution in the handmade papermaking from *Calotropis procera*." Pp. 77–88 *in* R. C. Kuhad & A. Singh (editors), *Biotechnology for Environmental Management and Resource Recovery.* Springer, New Delhi.

Dadswell, H. & A. J. Watson. 1961. "Influence of fibre morphology on paper properties." *Appita* 14: 168–178.

Dana, E. D., J. García-de-Lomas, F. Verloove, D. García-Ocaña, V. Gámez, J. Alcaraz & J. M. Ortiz. 2017. "*Colocasia esculenta* (L.) Schott (Araceae), an expanding invasive species of aquatic ecosystems in the Iberian Peninsula: New records and risk assessment." *Limnetica* 36: 15–27.

Drège, J. P. 1998. "First mentions of paper in Vietnam according to Chinese sources." French translation of "Những ghi chép dầu tien về làm giấy ở nghề nam qua thu tịch trung hoa." Việt Nam Học, Ký yếu Hội tháo quộc tế lấn thứ nhất, Hà Nội.

Dudek, K., M. Michlewicz, M. Dudek & P. Tryjanowski. 2016. "Invasive Canadian goldenrod (*Solidago canadensis* L.) as a preferred foraging habitat for spiders." *Arthropod-Plant Interactions* 10: 377–381.

Dutt, D., J. S. Upadhyaya, R. S. Malik & C. H. Tyagi. 2005. "Studies on the pulp and papermaking characteristics of some Indian non-woody fibrous raw materials. I." *Cellulose Chemistry and Technology* 39: 115–128.

Fanchette, S. 2016. "Papeterie et recyclage dans les villages de métier: La fin d'un modèle de production? (Delta du fleuve Rouge, Vietnam)." *Techniques & Culture* 65-66: 1–32.

Fanchette, S. & N. Stedman. 2009. *Discovering Craft Villages in Vietnam: Ten Itineraries Around Hà Nội*. Institut de recherche pour le développement, Hanoi.

Ferreira, P. J., J. A. Gamelas, M. G. Carvalho, G. V. Duarte, J. M. Canhoto & R. Passas. 2013. "Evaluation of the papermaking potential of *Ailanthus altissima*." *Industrial Crops and Products* 42: 538–542.

Forseth, I. N. & A. F. Innis. 2004. "Kudzu (*Pueraria montana*): History, physiology, and ecology combine to make a major ecosystem threat." *Critical Reviews in Plant Sciences* 23: 401–413.

Gary, J. 2012. "Paper mulberry trees, clumps, and oval beds: The first phase of landscape restoration at Thomas Jefferson's Poplar Forest." *Magnolia* 25(4): 1–11. https://southerngardenhistory.org/wp-content/uploads/2015/12/Magnolia_Fall_2012.pdf

Gentry, A. & R. Vazquez. 1993. *A Field Guide to the Families and Genera of Woody Plants of Northwest South America (Colombia, Ecuador, Peru), with Supplementary Notes on Herbaceous Taxa*. University of Chicago Press, Chicago.

Georgia Exotic Pest Plant Council. 2006. "List of non-native invasive plants in Georgia." https://www.se-eppc.org/wildlandweeds/pdf/fall2006-gaexoticslist-pp15-18.pdf

Ghersa, C. M., E. de la Fuente, S. Suarez & R. J. Leon. 2002. "Woody species invasion in the Rolling Pampa grasslands, Argentina." *Agriculture, Ecosystems & Environment* 88: 271–278.

Ghosh, S. R., D. C. Saikia, T. Goswami, B. P. Chaliha, J. N. Baruah, C. Effrem, D. D. Jatkar & G. Thyagarajan. 1984. "Utilization of water hyacinth (*Eichhornia crassipes*) for paper and board making." Pp. 436–460 *in* G. Thyagarajan (editor), *Proceedings of the International Conference on Water Hyacinth, Hyderabad, India*. Reports and Proceedings Series 7, United Nations Environment Programme, Nairobi.

Gordon, D. R., K. J. Tancig, D. A. Onderdonk & C. A. Gantz. 2011. "Assessing the invasive potential of biofuel species proposed for Florida and the United States using the Australian Weed Risk Assessment." *Biomass and Bioenergy* 35: 74–79.

Goswami, T. & C. N. Saikia. 1994. "Water hyacinth: A potential material for greaseproof paper." *Bioresource Technology* 50: 235–238.

Hamad, S. F., N. Stehling, S. A. Hayes, J. P. Foreman & C. Rodenburg. 2019. "Exploiting plasma exposed, natural surface nanostructures in ramie fibers for polymer composite applications." *Materials* 12: 1631.

Hark, M. & P. Boakye. 2022. "Adinkra-stamped handmade paper in Ntonso, Ghana." Pp. 2–9 *in* R. Pandey (editor), *Paper and Colour: Dyes and Dyeing around the World*. Legacy Press, Ann Arbor.

Helman-Ważny, A. 2014. *The Archaeology of Tibetan Books*. Brill, Leiden and Boston.

Hending, D., H. Randrianarison, M. Holderied, G. McCabe & S. Cotton. 2021. "The kapok tree (*Ceiba pentandra* (L.) Gaertn, Malvaceae) as a food source for native vertebrate species during times of resource scarcity and its potential for reforestation in Madagascar." *Austral Ecology* 46: 1440–1444.

Hiebert, H. 2006. *Papermaking with Garden Plants and Common Weeds*. Storey Publishing, North Adams, Massachusetts.

Holmes, W. C., J. F. Pruski & J. R. Singhurst. 2000. "*Thymelaea passerina* (Thymelaeaceae) new to Texas." *SIDA Contributions to Botany* 19: 403–406.

Ilvessalo-Pfäffli, M.-S. 1995. *Fiber Atlas: Identification of Papermaking Fibers*. Springer Berlin, Heidelberg.

Imaeda, Y. 1989. "Papermaking in Bhutan." *Acta Orientalia Academiae Scientiarum Hungaricae* 43: 409–414.

Isenberg, I. H. 1956. "Papermaking fibers." *Economic Botany* 10: 176–193.

Jewett, D. K., C. J. Jiang, K. O. Britton, J. H. Sun & J. Tang. 2003. "Characterizing specimens of kudzu and related taxa with RAPD's." *Castanea* 68: 254–260.

Joedodibroto, R., L. S. Widyanto & M. Soerjani. 1983. "Potential uses of some aquatic weeds as paper pulp." *Journal of Aquatic Plant Management* 21: 29–32.

Johnson, J. 2010. "Paper sample: Comparing Japanese knotweed fibers harvested three weeks apart." *Hand Papermaking* 25(1): 16–19.

Kiaei, M., A. Samariha & J. E. Kasmani. 2011. "Characterization of biometry and the chemical and morphological properties of fibers from bagasse, corn, sunflower, rice and rapeseed residues in Iran." *African Journal of Agricultural Research* 6: 3762–3767.

Kim, C.-H., J.-Y. Lee, H.-J. Gwak, H.-J. Lee, K. K. Back, J.-M. Seo & H.-J. Park. 2010. "Study on the properties of kudzu fibers as a papermaking material." *Journal of Korea Technical Association of The Pulp and Paper Industry* 42: 53–60.

Kitajima, S., K. Kamei, S. Taketani, M. Yamaguchi, F. Kawai, A. Komatsu & Y. Inukai. 2010. "Two chitinase-like proteins abundantly accumulated in latex of mulberry show insecticidal activity." *BMC Biochemistry* 11: 1–7.

Kostel, G. 2009. "*Thymelaea passerina* (Thymelaeaceae) in South Dakota." *Journal of the Botanical Research Institute of Texas* 3: 901–903.

Krauss, U. 2012. "Invasive alien species management in St. Lucia and Caribbean partner countries." Pp. 196–206 *in* J.-L. Vernier & M. Burac (editors), *Biodiversité insulaire: la flore, la faune et l'homme dans les Petites Antilles*. Actes du Colloque international 1–79. Université Antilles-Guyane, Schœlcher, Martinique.

Kuo, W.-H., S.-H. Liu, C.-C. Chang, C.-L. Hsieh, Y.-H. Li, T. Ito, H. Won, G. Kokubugata & K.-F. Chung. 2022. "Plastome phylogenomics of *Allaeanthus*, *Broussonetia* and *Malaisia* (Dorstenieae, Moraceae) and the origin of *B.* ×*kazinoki*." *Journal of Plant Research* 135: 203–220.

Laroque, C. 2020. "Tonkin's giấy dó and its Chinese roots." *Z Badań nad Książką i Księgozbiorami Historycznymi* 14: 451–487.

Lee, A. 2012. *Hanji Unfurled*. Legacy Press, Ann Arbor.

Lee, A. 2022. "Deliberate collaboration with Hyeyung." Blog post, 25 November 2022. http://moonaimee.blogspot.com/2022/11/deliberate-collaboration-with-hyeyung.html

Lee, J.-Y., C.-H. Kim, J.-E. Lee, S. Kwon, H.-H. Park, H.-T. Yim, S.-O. Moon & H.-G. Gu. 2018. "Papermaking characteristics of kudzu root fibers with different morphology." *Journal of Korea TAPPI* 50(5): 22–30.

Li, H., H. Sun, L. Pu & Z. He. 2014. "*Stellera chamaejasme* roots as raw material for pulp production." *BioResources* 9(3): 4775–4783.

Li, T. 2018. "Identifying sources of fibre in Chinese handmade papers by phytoliths: A methodological exploration." *Science and Technology of Archaeological Research* 4: 1–11.

Li, X., J. Guo & B. Wang. 2015. "A study of ancient paper fragments from an Eastern Han Dynasty tomb in Minfeng County, Xinjiang Uygur Autonomous Region." *Chinese Cultural Relics* 2: 366–370.

Little, Jr., E. L. & R. G. Skolmen. 1989. "Gunpowder-tree." *Common Forest Trees of Hawaii (Native and Introduced)*. Agricultural Bulletin no. 679. U.S. Forest Service, U.S. Department of Agriculture, Washington, D.C. Reprinted by College of Tropical Agriculture and Human Resources, University of Hawai'i at Mānoa, 2003. https://www.ctahr.hawaii.edu/gsp/doc/Forestry/Little_Skolmen_CFT/CFT_Trema_orientalis.pdf

López Binnqüist, C., A. Quintanar–Isaías & M. Vander Meeren. 2012. "Mexican bark paper: Evidence of history of tree species used and their fiber characteristics." *Economic Botany* 66: 138–148.

Lutz, W. 1983. "Non-Japanese fibers for Japanese papermaking." Appendix One *in* T. Barrett, *Japanese Papermaking: Traditions, Tools, and Techniques*. Weatherhill, New York, Tokyo.

Maideliza, T., R. R. D. Mayerni & D. Rezki. 2017. "Comparative study of length and growth rate of ramie (*Boehmeria nivea* L. Gaud.) bast fiber of Indonesian clones." *International Journal of Advanced Science, Engineering, Information, and Technology* 7: 2273–2278.

Main, N. M., R. A. Talib, A. Z. Mohamed, R. A. Rahman & R. Ibrahim. 2014. "Suitability of coir fibers as pulp and paper." *Agriculture and Agricultural Science Procedia* 2: 304–311. https://doi.org/10.1016/j.aaspro.2014.11.043

Maji, S., R. Mehrotra & S. Mehrotra. 2013. "Extraction of high quality cellulose from the stem of *Calotropis procera*." *South Asian Journal of Experimental Biology* 3: 113–118.

Marques, G., J. Rencoret, A. Gutiérrez, J. E. M. Alfonso & J. C. del Río. 2010. "Evaluation of the chemical composition of different non-woody plant fibers used for pulp and paper manufacturing." *The Open Agriculture Journal* 4: 93–101.

Marshall, B. G. 2006. "Fiber and community upliftment in South Africa." *Hand Papermaking* 21(2): 14–16.

McLaughlin, S. P. 2000. "Properties of paper made from fibers of *Hesperaloe funifera* (Agavaceae)." *Economic Botany* 54: 192–196.

McLaughlin, S. P. & S. M. Schuck. 1991. "Fiber properties of several species of Agavaceae from the southwestern United States and northern Mexico." *Economic Botany* 45: 480–486.

Mendes-Rodrigues, C., P. E. Oliveira, R. C. Marinho, R. Romero & M. A. Ranal. 2019. "Are the alien species of Melastomataceae and Bombacoideae a potential risk for Brazilian Cerrado?" *Open Access Library Journal* 6: 1–4.

Miller, S. 2010. "Making paper from the banana plant: An Alabama approach." *Hand Papermaking* 25(1): 33–35.

Mitich, L. W. 2000. "Kudzu [*Pueraria lobata* (Willd.) Ohwi]." *Weed Technology* 14: 231–235.

Morgan, E. C., W. A. Overholt & B. Sellers. 2004. "Wildland weeds: Paper mulberry, *Broussonetia papyrifera*." University of Florida IFAS Extension. https://www.growables.org/information/TropicalFruit/documents/MulberryPaperEDIS.pdf

Mullock, H. 1995. "Xuan paper." *The Paper Conservator* 19: 23–30.

Ojascastro, J. 2023. *Following the Paper Trail: An Ethnoecology of Hand Papermaking Traditions*. Ph.D. thesis, Washington University in St. Louis. https://www.proquest.com/docview/2901859563

Ojascastro, J., V. Phạm, H. N. Trân & R. Hart. 2024. *Hand Papermaking Traditions of Việt Nam*. Society of Ethnobiology.

Olotuah, O. F. 2006. "Suitability of some local bast fibre plants in pulp and paper making." *Journal of Biological Sciences* 6: 635–637.

Pang, B. 1992. *Identification of Plant Fibers in Hawaiian Kapa: From Ethnology to Botany*. Graduate thesis, University of Hawai'i at Mānoa.

Peña-Ahumada, B., M. Saldarriaga-Córdoba, O. Kardailsky, X. Moncada, M. Moraga, E. Matisoo-Smith, D. Seelenfreund & A. Seelenfreund. 2020. "A tale of textiles: Genetic characterization of historical paper mulberry barkcloth from Oceania." *PLoS ONE* 15(5): 1–20. https://doi.org/10.1371/journal.pone.0233113

Peñailillo, J., G. Olivares, X. Moncada, C. Payacán, C. S. Chang, K. F. Chung, P. J. Matthews, A. Seelenfreund & D. Seelenfreund. 2016. "Sex distribution of Paper Mulberry (Broussonetia papyrifera) in the Pacific." *PLoS ONE* 11(8): 1–19. https://doi.org/10.1371/journal.pone.0163188

Peters, C. M., J. Rosenthal & T. Urbina. 1987. "Otomi bark paper in Mexico: Commercialization of a pre-hispanic technology." *Economic Botany* 41: 423–432.

Petty, G. 2010. "Journey into papermaking: A focus on Kentucky invasive plants." *Hand Papermaking* 25(1): 3–11.

Pohl, R. W. 1955. "*Thymelaea passerina*, a new weed in the United States." *Proceedings of the Iowa Academy of Science* 62: 152–154.

Puran, B. & S. B. Ronell. 2014. "Invasive weed risk assessment of three potential bioenergy fuel species." *International Journal of Biodiversity and Conservation* 6: 790–796.

Rantoandro, G. 1983. "Contribution à la connaissance du « papier Antemoro » (Sud-est de Madagascar)." *Archipel* 26: 86–104.

Reddy, N. & Y. Yang. 2008. "Characterizing natural cellulose fibers from velvet leaf (*Abutilon theophrasti*) stems." *Bioresource Technology* 99: 2449–2454.

Rich, J. 2022. "Milkweed fluff paper." https://www.instagram.com/p/Cghe4qSLKMX/?hl=en

Richard, A. 2010. "A papermaker's dilemma: Examining the use of invasive plants." *Hand Papermaking* 25(1): 23–27.

Rickleton, C. 2018. "The Uzbek entrepreneur tapping paper's age-old power." Phys.org. https://phys.org/news/2018-06-uzbek-entrepreneur-paper-age-old-power.html

Rivers, R. 2020. "Nettle paper, 21/06/20." Web. https://cpb-eu-w2.wpmucdn.com/blogs.ucl.ac.uk/dist/b/306/files/2020/07/Robert-Rivers-Nettle-Paper.pdf

Rowell, R. M., J. S. Han & J. S. Rowell. 2000. "Characterization and factors effecting fiber properties." Pp. 115–134 *in* E. Frollini, A. Leao & L. H. Capparelli Mattoso (editors), *Natural Polymers and Agrofibers Based Composites*. Embrapa Instrumentacão Agropeçu, Sao Carlos.

Saeed, H. A., Y. Liu, L. A. Lucia & H. Chen. 2017. "Evaluation of Sudanese sorghum and bagasse as a pulp and paper feedstock." *BioResources* 12: 5212–5222.

Saeed, H., Y. Liu, L. Lucia & H. Chen. 2018. "Sudanese dicots as alternative fiber sources for pulp and papermaking." *Drvna Industrija* 69: 175–182.

Sam-Brew, S. & G. D. Smith. 2017. "Flax shive and hemp hurd residues as alternative raw material for particleboard production." *BioResources* 12: 5715–5735.

Schäffer, H. C. 1765. *Versuche und Muster ohne alle Lumpen oder doch mit enem geringen Zusatze derselben Papier zu machen*. Regensburg.

Schattenburg-Raymond, L. 2020. "A new perspective on Hawaiian kapa-making." Pp. 73–81 *in* F. Lennard & A. Mills (editors), *Material Approaches to Polynesian Barkcloth: Cloth, Connections, Communities*. Sidestone Press, Leiden.

Schmidt, J. & N. Stavisky. 1983. "Uses of *Thymelaea hirsuta* (Mitnan) with emphasis on hand papermaking." *Economic Botany* 37: 310–321.

Sharma, M., C. L. Sharma & Y. B. Kumar. 2013. "Evaluation of fiber characteristics in some weeds of Arunachal Pradesh, India for pulp and paper making." *Research Journal of Agriculture and Forestry Sciences* 1(3): 15–21.

Sheahan, C. M. 2012. "USDA Plant Guide for sunn hemp (*Crotalaria juncea*)." United States Department of Agriculture. https://plants.usda.gov/DocumentLibrary/plantguide/pdf/pg_crju.pdf

Singleton, M. 2012. *Eight Thousand Daughters Woven Into Bayou Braids*. MFA thesis dissertation, Louisiana State University, Baton Rouge. https://www.megansingleton.com/eight-thousand-daughters.html

Szatmari, P. M. 2012. "Alien and invasive plants in Carei Plain Natural Protected Area, Western Romania: Impact on natural habitats and conservation implications." *Biology and Environment* 3: 109–120.

Tasillo, M. 2010. "Guerrilla weeding and the practice of slowing down." *Hand Papermaking* 25(1): 12–15.

Terry, N. 2022. "Agua papel." Origami-shop.com. https://www.origami-shop.com/en/papers-agua-papels-fabian-correa-xsl-207_215_2515_2260_2265.html

Thomas, P. & D. Thomas. 1999. *Paper from Plants*. Peter & Donna Thomas, Santa Cruz.

Trier, J. 1972. *Ancient Paper of Nepal: Results of Ethno-technological Fieldwork on its Manufacture, Uses, and History—With Technical Analyses of Bast, Paper, and Manuscripts*. Jutland Archaeological Society Publications, Copenhagen.

Trusty, J. L., L. R. Goertzen, W. C. Zipperer & B. G. Lockaby. 2007. "Invasive *Wisteria* in the Southeastern United States: Genetic diversity, hybridization and the role of urban centers." *Urban Ecosystems* 10: 379–395.

Tutus, A., N. Çömlekcioğlu, S. Karaman & M. H. Alma. 2010. "Chemical composition and fiber properties of *Crambe orientalis* and *C. tataria*." *International Journal of Agriculture and Biology* 12: 286–290.

University of Florida. 2022. "*Ficus microcarpa*." University of Florida Center for Aquatic and Invasive Plants. https://plants.ifas.ufl.edu/plant-directory/ficus-microcarpa/#, accessed 20 September 2022.

Ververis, C., K. Georghiou, N. Christodoulakis, P. Santas & R. Santas. 2004. "Fiber dimensions, lignin and cellulose content of various plant materials and their suitability for paper production." *Industrial Crops and Products* 19: 245–254.

Vincent, M. A. & J. W. Thieret. 1987. "*Thymelaea passerina* (Thymelaeaceae) in Ohio." *SIDA, Contributions to Botany* 12: 75–78.

von Hagen, V. W. 1944. *The Aztec and Maya Papermakers*. J. J. Augustin Publisher, New York.

Walker, G. 1927. "The Garo manufacture of bark cloth." *Man* 27: 15.

Washington State Noxious Weed Control Board. 2006. "Written findings of the Washington State Noxious Weed Control Board (September 2006): *Daphne laureola* L." https://www.nwcb.wa.gov/images/weeds/Daphne_laureola_Written-Findings.pdf

Weeks, L. 2011. "The art of war on invasive species." *NPR*, 28 February 2011. https://www.npr.org/2011/02/28/134054004/the-art-of-war-on-invasive-species

White, G. A. & J. R. Haun. 1965. "Growing *Crotalaria juncea*, a multi-purpose legume, for paper pulp." *Economic Botany* 19: 175–183.

Yao, N. & S. Wei. 2021. "Characterization and identification of traditional Chinese handmade paper via pyrolysis-gas chromatography-mass spectrometry." *BioResources* 16: 3942–3951.

Yum, H. 2008. *Traditional Korean Papermaking: History, Techniques, and Materials*. Doctoral thesis, University of Northumbria, Newcastle.

Zerega, N. J., D. Ragone & T. J. Motley. 2004. "Complex origins of breadfruit (*Artocarpus altilis*, Moraceae): Implications for human migrations in Oceania." *American Journal of Botany* 91: 760–766.

TABLE 1. Invasive plants that have been used in experimental hand papermaking. Species in **bold** yield papers of exceptionally high quality.

Common name	Species	Botanical family	Tissue used	Citation
Alligator weed	*Alternanthera philoxeroides*	Amaranthaceae	Leaf; stem	Singleton, 2012
Queen Anne's lace	*Daucus carota*	Apiaceae	Stem, flowers, seeds	Castañuela, 2010
Taro	*Colocasia esculenta*	Araceae	Leaf	Mowers, pers. comm.
English ivy	*Hedera helix*	Araliaceae	Stem	Clark, in Johnson, 2010; Clark, in Weeks, 2011
Lesser burdock	*Arctium minus*	Asteraceae	Stem	Schäffer, 1765; Bell, 1995
Mugwort	*Artemisia vulgaris*	Asteraceae	Stem	Silverberg & St. Germain, in Thomas & Thomas, 1999
Nodding or musk thistle	*Carduus nutans*	Asteraceae	Seed hairs	Petty, 2010
Bull thistle	*Cirsium vulgare*	Asteraceae	Seed hairs	Schäffer, 1765; Bell, 1995
Garlic mustard	*Alliaria petiolata*	Brassicaceae	Stem	Clark, in Weeks, 2018
Common hop	*Humulus lupulus*	Cannabaceae	Phloem	Isenberg, 1956
Amur honeysuckle	*Lonicera maackii*	Caprifoliaceae	Phloem	Heeres, in Babcock, 2017
Chinese bittersweet	*Celastrus orbiculatus*	Celastraceae	Stem	Clark, in Petty, 2010
Burningbush	*Euonymus alatus*	Celastraceae	Phloem	Lutz, 1983
Purple nutsedge	*Cyperus rotundus*	Cyperaceae	Stem	Bidin et al., 2015
Castor bean	*Ricinus communis*	Euphorbiaceae	Phloem	Torres & Valenzuela, pers. comm.
Sunn hemp	***Crotalaria juncea***	Fabaceae	Phloem	White & Haun, 1965
Scotch broom	*Cytisus scoparius*	Fabaceae	Phloem	Bell, 1995

continued

TABLE 1. *(continued)*

Common name	Species	Botanical family	Tissue used	Citation
Kudzu	*Pueraria montana* var. *lobata*	Fabaceae	Phloem	Petty, 2010
Chinese wisteria	*Wisteria sinensis*	Fabaceae	Phloem	Bell, 1995
Japanese wisteria	*Wisteria floribunda*	Fabaceae	Phloem	Lutz, 1983; Bell, 1995
Hybrid wisteria	*Wisteria ×formosa*	Fabaceae	Phloem	Bell, 1995; Ojascastro, 2023
Yellow flag	*Iris pseudacorus*	Iridaceae	Leaf	Johnson, 2010; Singel, in Cappiello, 2021
Velvetleaf	*Abutilon theophrasti*	Malvaceae	Phloem	Bell, 1995
Cuban jute	*Sida rhombifolia*	Malvaceae	Phloem	Bell, 1995
Caesarweed	*Urena lobata*	Malvaceae	Phloem	Bell, 1995
Paper mulberry	***Broussonetia papyrifera***	Moraceae	Phloem	Richard, 2010; Ojascastro, 2023
White mulberry	***Morus alba***	Moraceae	Phloem	Lutz, 1983; Petty, 2010; Ojascastro, 2023
Princesstree	*Paulownia tomentosa*	Paulowniaceae	Leaf (petiole)	Tasillo, 2010
Reed canary grass	*Phalaris arundinacea*	Poaceae	Stem	Johnson, in Thomas & Thomas, 1999
Giant reed	*Arundo donax*	Poaceae	Stem	Marques et al., 2010
Pampas grass	*Cortaderia selloana*	Poaceae	Stem	Bell, 1995; Thomas & Thomas, 1999
Common reed	*Phragmites australis*	Poaceae	Stem	Bell, 1995
Japanese knotweed	*Reynoutria japonica*	Polygonaceae	Stem	Johnson, 2010
Curly dock	*Rumex crispus*	Polygonaceae	Leaf (petiole)	Johnson, 2010; Petty, 2010
Water hyacinth	*Pontederia crassipes*	Pontederiaceae	Leaf	Goswami & Saikia, 1994; Singleton, 2012

Common name	Species	Botanical family	Tissue used	Citation
Multiflora rose	*Rosa multiflora*	Rosaceae	Stem	Clark, in Weeks, 2011
Tree of heaven	*Ailanthus altissima*	Simaroubaceae	Phloem	Petty, 2010; Singel, in Cappiello, 2021
Spurge-laurel	***Daphne laureola***	Thymelaeaceae	Phloem	Bonham, in Thomas & Thomas, 1999
Mezereon	***Daphne mezereum***	Thymelaeaceae	Phloem	Lutz, 1983; Ojascastro, 2023
Siberian elm	*Ulmus pumila*	Ulmaceae	Phloem	Lutz, 1983
Stinging nettle	*Urtica dioica*	Urticaceae	Phloem	Beisinghoff, 2010; Rivers, 2020
White ginger	*Hedychium coronarium*	Zingiberaceae	Stem	Torres & Valenzuela, pers. comm.

TABLE 2. Recommended invasive species to try for hand papermaking.

Common name	Species	Botanical family	Justification
Tall goldenrod	*Solidago altissima*	Asteraceae	Hairs on wind-dispersed fruits could be leveraged for paper
Common dandelion	*Taraxacum officinale*	Asteraceae	Hairs on wind-dispersed fruits could be leveraged for paper
Japanese hops	*Humulus japonicus*	Cannabaceae	Congener of common hop (*H. lupulus*), used off and on for paper, twine, and cordage
Gunpowder tree	*Trema orientale*	Cannabaceae	Congener of jonote colorado (*T. micranthum*), used to make amate in Mexico
Chinese tallowtree	*Triadica sebifera*	Euphorbiaceae	Formerly *Sapium sebiferum*; therefore a close relative of jonote brujo (*S. glandulosum*), used to make amate for brujería in Mexico
Chinese banyan	*Ficus microcarpa*	Moraceae	Congener of xalama (*Ficus aurea*), once used to make amate in Mexico
Annual thymelaea	*Thymelaea passerina*	Thymelaeaceae	Congener of mitnan (*T. hirsuta*), used to make paper in the Levant

Contributors ◆ ◆ ◆

Wendy L. Applequist, an associate scientist in the Missouri Botanical Garden's William L. Brown Center, is a plant taxonomist with interests in medicinal plants, nomenclature, and the flora of Madagascar.

Franziska Eller is a biologist with experience in plant eco-physiology and an interest in vegetation responses to global change. Her current position is as senior specialist in climate at SEGES Innovation, a nonprofit, independent agricultural research- and innovation company located in Denmark.

Alexia Franzidis is an associate professor of tourism, recreation, and sport at the University of North Carolina Wilmington; her research focuses on sustainability in leisure and tourism.

Joshua Ulan Galperin is an associate professor of law at the Elisabeth Haub School of Law at Pace University; his scholarship focuses on environmental law, administrative law, food law, and democratic governance.

Thomas Avery Garran is the president of Herb Whisperer, Inc., and executive director of the East West Herb School. An independent scholar, author, and instructor with interests that include Chinese herbal medicine, medicinal plant processing and pharmacy, and medicinal plant agriculture. He lives in western Massachusetts where he and his wife steward the land, creating a botanical sanctuary and learning center.

Katie Grove is an educator, artist, and leader in the field of natural basketry. She teaches wild basketry classes with a focus on ethical harvesting and nature connection full time from her studio in upstate New York.

Theresa Hornstein taught biology at Lake Superior College for 35 years. Now retired, she continues working as a fiber artist and teaching at folk schools throughout Minnesota and Wisconsin.

Mohammad Anwar Hossain is a professor in the Department of Genetics and Plant Breeding at Bangladesh Agricultural University in Mymensingh, Bangladesh.

Katie Carter King is a freelance writer, researcher, and editor who specializes in using all-too-often overlooked history to tell important, engaging stories that challenge people's notions of the past and change their perception of the present. A child of the Georgia Piedmont, she currently lives in San Francisco.

Sara E. Kuebbing is the Director of Research of the Yale Applied Science Synthesis Program at The Forest School at the Yale School of the Environment and is a restoration ecologist interested in managing ecosystems to support biodiversity and ecosystem services.

Katie MacDonald pioneers new biomaterial assemblies as assistant professor of architecture at the University of Virginia, director of the Before Building Laboratory, and cofounder of After Architecture.

Peter J. Matthews is professor at the National Museum of Ethnology, Osaka, Japan, where he works in the fields of crop history, ethnobotany, and prehistory, in collaboration with counterparts across Asia and the Pacific.

Van Dzu Nguyen is associate professor at the Institute of Ecology and Biological Resources, Hanoi, Vietnam, where he works as a plant taxonomist and ethnobotanist; he is working on plant conservation in collaboration with the Atlanta Botanical Garden, Royal Botanic Gardens, Kew, and Royal Botanic Garden Edinburgh.

Martin A. Nuñez, an associate professor at the University of Houston, is an ecologist mainly focused on biological invasions.

James Ojascastro is a U.S.-based ethnobotanist, papermaker, and origamist who uses scientific methods to inform and inspire his art. He completed his Ph.D. in 2023 at Washington University in St. Louis, where he studied the ecological connections between plants and people through the lens of artisanal papermaking.

Kyle Schumann advances the accessibility of robotic construction technology as assistant professor of architecture at the University of Virginia, director of the Before Building Laboratory, and cofounder of After Architecture.

Alana N. Seaman is an associate professor of tourism, recreation, and sport at the University of North Carolina Wilmington; her research examines cultural facets of food, sports, and other trends in hospitality, tourism, and leisure.

Kathy Voth is the founder and editor of OnPasture.com, a free, online library of articles translating science and experience into practices graziers can use to be more sustainable and profitable. She invented the method for teaching livestock to add invasive weeds to their diets as part of her goal of improving the lives of graziers.

Tusha Yakovleva works at the Center for Native Peoples and the Environment (SUNY College of Environmental Science and Forestry) in support of reciprocal relationships between land and people. Tusha's ethnobotanical writing can be found at www.foundwith.care.